RECOGNISING ACHIEVEMENT

Official Publi

CW00343236

OCR DESIGN & TECHNOLOGY FOR GCSE

TEXTILES TECHNOLOGY

JAYNE MARCH

MARIA JAMES

CAREY CLARKSON

EDITOR: BOB WHITE

HODDER
EDUCATION
AN HACHETTE UK COMPANY

Orders: please contact Bookpoint Ltd, 130 Milton Park, Abingdon, Oxon
OX14 4SB. Telephone: +44 (0)1235 8 27720. Fax: +44 (0)1235 400454.
Lines are open from 9.00am to 5.00pm, Monday to Saturday, with a 24-
hour message-answering service. You can also order through our website
www.hoddereducation.co.uk

If you have any comments to make about this, or any of our other titles,
please send them to educationenquiries@hodder.co.uk

British Library Cataloguing in Publication Data
A catalogue record for this title is available from the British Library

ISBN: 978 0 340 98199 3

First Edition Published 2009
mpression number   10 9 8 7 6 5 4 3 2 1
Year   2012 2011 2010 2009

Hachette UK's policy is to use papers that are natural, renewable and
recyclable products and made from wood grown in sustainable forests.
The logging and manufacturing processes are expected to conform to the
environmental regulations of the country of origin.

Cover photo from Don Farrall/Photodisc
Typeset by Fakenham Photosetting Ltd, Fakenham, Norfolk
Printed in Italy for Hodder Education, an Hachette UK Company,
338 Euston Road, London NW1 3BH

CONTENTS

CHAPTER 12 UNIT A574: TECHNICAL ASPECTS OF DESIGNING AND MAKING — 195

HOW TO GET THE MOST OUT OF THIS BOOK

Welcome to OCR Design and Technology for GCSE Textiles Technology (specification numbers J307 and J047).

The book has been designed to support you throughout your GCSE course. It provides clear and precise guidance for each of the four units that make up the full course qualification, along with detailed information about the subject content of the course. It will be an extremely effective resource in helping you prepare for both controlled assessment and examined units.

The book has been written and developed by a team of writers who all have considerable specialist knowledge of the subject area and are all very experienced teachers.

The book:
- *is student focused. The aim of the book is to help you achieve the best possible results from your study of GCSE Textiles Technology*
- *gives clear guidance of exactly what is expected of you in both controlled assessment and examined units*
- *contains examiner tips and guidance to help improve your performance in both Controlled Assessment and examined units*
- *provides detailed information relating to the subject content and designing*
- *is designed to help you locate information quickly*
- *is focused on the OCR specification for GCSE Textiles Technology*
- *has relevance and value to other GCSE Textiles Technology courses.*

The book outlines the knowledge, skills and understanding required to be successful within GCSE Textiles Technology. It is designed to give you a 'body of knowledge' which can be used to develop your own knowledge and understanding during the course and support you when undertaking both controlled assessment and examined units.

Chapters 1–8 form the 'body of knowledge'. Chapters 9–12 give specific guidance about each of the units that make up the GCSE course.

▶ Unit A571 Introduction to Designing and Making

Chapter 9 gives detailed information about the structure of the controlled assessment unit and the rules relating to the controlled assessment task you will undertake. It clearly explains what you need to do section by section and includes examiner tips to help improve your performance. Specific reference is made to the assessment criteria and an explanation is provided as to how the criteria will be applied to your product. Examples of students' work are used within the text to reinforce the requirements of each section.

▶ Unit A572 Sustainable Design

This chapter provides detailed information relating to this unit. It gives a clear explanation of the structure of the examination and gives further information relating to the key aspects of sustainability in relation to GCSE Textiles Technology. The chapter examines:

- The 6R's in relation to Textiles Technology
- The social and moral issues linked to the manufacture, use and disposal of products manufactered using Textiles Technology
- The impact of cultural issues on Textiles
- How to select materials that are both suitable and sustainable
- Current issues affecting the design of new products.

▶ Unit A573 Making Quality Products

Chapter 11 follows a similar format to Chapter 9. It explains the requirements of the unit section by section and includes examiner tips to guide you through the controlled assessment task.

▶ Unit A574 Technical Aspects of Designing and Making

Chapter 12 is designed to help you prepare for the written examination. It clearly describes the format of the examination paper and gives examples of questions. Examiner tips are given to help you identify the type of question and the approach you should take in completing your answer.

▶ Icons used in this book

Introduction boxes provide a short overview of the topics under discussion in the section.

KEY POINTS

- Key Points boxes list key aspects of a topic.

KEY TERMS

Key Terms boxes provide definitions of the technical terms used in the section.

EXAMINER'S TIPS

Examiner's Tips boxes give tips on how to improve performance in both the Controlled Assessment and examined units.

LEARNING OUTCOMES

Learning Outcomes boxes highlight the knowledge and understanding you should have developed by the end of the section.

ACTIVITY

Activity boxes suggest interesting tasks to support, enhance and extend learning opportunities.

CASE STUDY

Case study boxes provide examples of how real-life businesses use the knowledge and skills discussed.

QUESTIONS

Questions boxes provide practice questions to test key areas of the content of the specification.

All examples of student work included in this book are available in larger format on the OCD Design & Technology for GCSE: Textiles Technology Teacher's Resource DVD-ROM (ISBN 978 0 340 91190).

ACKNOWLEDGEMENTS

The authors would like to thank the following: Maralyn Butler, Sarah Fox, Nigel Harris, Phoebe James and Alex Russell. Special thanks go to the following centres that have provided some of the example GCSE coursework pieces throughout the book: Queen Elizabeth's Grammar School, Hemel Hempstead School and Wasley Hills High School. Thanks also go to Peter from Hartlepool Sewing Centre, Teeside, who provided up-to-date infromation about the latest sewing machine and software sensations.

The authors and publishers would like to thank the following for use of photographs and illustrations in this volume:

Figure 1.1 © Daniel Kourey/iStockphoto.com; Figure 1.2 © Grigory Bibikov/ iStockphoto.com; Figures 1.6–1.8 with kind permission by Alex Russell; Figure 1.9 © LaModa/Alamy; Figures 1.10–1.13 with kind permission by Alex Russell; Figure 1.15 Bernina My Label images © Bernina International AG, www.bernina.co.uk; Figure 1.16 Bernina My Label images © Bernina International AG, www.bernina.co.uk; 2.1 © Frank van den Bergh/iStockphoto.com; Figure 2.2 © B. Sieckmann/iStockphoto.com; Figure 2.3 © Duncan Babbage/iStockphoto.com; Figure 2.4 © Sieto Verver/iStockphoto.com; Figure 2.5 © Maryann Frazier/Science Photo Library; Figure 2.7 © Shaun Lowe/istockphoto.com; Figure 2.8 © Jim DeLillo/istockphoto.com; 2.12 © Yasain GUNEYSU/istockphoto.com; Figure 2.34 © Tim McCaig/iStockphoto.com; 2.35 © Archimage/Alamy; Figure 2.36 © Jacob Wackerhausen/iStockphoto.com; Figure 2.44 © Micha Adamczyk/iStockphoto.com; Figure 2.45 © Jason Maehl/iStockphoto.com; Figure 2.46 © Sasha Radosavljevic/iStockphoto.com; 2.47 Reproduced with kind permission by W. L. Gore & Associates; Figure 2.49 © Philips; Figure 2.50 with kind permission by ARC Outdoors; Figure 2.51 Fabrican/Gene Kiegel Photo by Gene Kiegel, www.genekiegel.com; Figure 3.1 © iStockphoto.com; Figures 3.4a and 3.4b all images of Bernina Sewing Machines © Bernina International AG, www.bernina.co.uk; Figure 3.5 © iStockphoto.com; Figure 3.44 © iStockphoto.com; Figure 3.45 © iStockphoto.com; Figure 3.53 © iStockphoto.com, © Trista Weibell iStockphoto.com, © Ryan Lane/ iStockphoto.com; Figure 3.61 © Kevin Schafer/Alamy; Figure 4.3 Jack Carey/Alamy Carey; Figure 4.4 © Stephen Green/iStockphoto.com; 4.6 © MIRALab-Universitaet Genf; Figure 4.8 © Youssouf Cader/iStockphoto.com; Figure 4.9 © VSM Group AB, with permission; Figure 4.10 © Philippe Plailly/Science Photo Library; Figure

4.11 © ITAR TASS/Bandphoto/Uppa Photoshot; Figure 5.1 © Eric Hood/iStockphoto.com; Figure 5.2 © iStockphoto.com; 5.4 © iStockphoto.com; Figure 5.6 © Geoffrey Hammond/iStockphoto.com; Figure 5.9 with kind permission by Mark Liu; Figure 5.10 © SERDAR YAGCI/iStockphoto.com; Figure 5.11 © Martha Bayona/iStockphoto.com; Figure 5.12 © iStockphoto.com; Figure 5.15 with kind permission by Avery Dennison; Figure 5.16 © iStockphoto.com; Figure 5.17 with kind permission by Icebreaker New Zealand Ltd; Figure 5.25 © Intertek; Figure 6.1 © iStockphoto.com; Figure 6.2 © Ljupco Smokovski/iStockphoto.com; Figure 6.3 © Skip ODonnell/iStockphoto.com; Figure 6.5 © Ray Tang/Rex Features; Figure 6.6 taken from *The Sustainability Handbook*, © ITDG Publishing and printed with kind permission by Practical Action; Figure 6.7 © 2005 Rob Crandall/The Image Works/Topfoto; Figure 6.8 with kind permission by So Indigo; Figures Figure 6.9 and 6.10 with kind permission by Sara Simmonds and Sharkah Chakra; Figures 7.2 and 7.3 © Danish Khan/iStockphoto.com; Figure 7.6 © Danish Khan/iStockphoto.com; Figure 9.1 © iStockphoto.com; Figure 10.3 © Tye Carnelli/iStockphoto.com; Figure 10.4 © Achim Prill/iStockphoto.com; Figure 10.5 © Marcus Clackson/iStockphoto.com; Figure 10.6 © Olivier Blondeau/iStockphoto.com; 10.7 © Gary Unwin – Fotolia.com; Figure 10.8 © Dawn Hudson – Fotolia.com; Figure 10.9 reproduced with permission of the Faitrade Foundation; Figure 10.10 © Dena Steiner/iStockphoto.com; Figure 11.1 © Gina Smith/iStockphoto.com

All other photos in this volume taken by the authors.
Illustrations by Art Construction.

Every effort has been made to trace and acknowledge ownership of copyright. The publishers will be happy to make arrangements with any copyright owners that it has not been possible to contact.

DESIGNING AND PRODUCTION PLANNING

By the end of this section you should have developed a knowledge and understanding of:

- How to work to a brief
- Analysing and evaluating a product
- Creating a design specification using your brief and your research
- The principles of good design and how to use them in your own work
- Other designers' work and their influence
- The use of digital media and new technologies in designing
- The process of designing a product and the tools you can use to help you
- How to plan the making of a product and what you need to consider in your planning, to aid production

1.1 WORKING TO A BRIEF

In this section you will learn about:

- developing a design brief
- producing an appropriate and considered response to a design brief.

*Ideas for new textile items do not just happen! The majority are based upon existing products which are redesigned to respond to current trends. In all cases, a **need** has been identified, usually through detailed research.*

Developing a design brief

In order to produce an appropriate design brief on which to base your portfolio you will need to be able to respond to a chosen theme in both of your controlled assessment units (A571 and A573). The **design brief** usually outlines the item to be designed, its end use and the user.

The starting point for your controlled assessment portfolio can take the form of a theme or a specific product associated with the theme. For example, if the chosen theme is 'Eco-wear' you may decide to design and model a textile item using fabric remnants and parts of existing products, or identify a pair of old jeans as your starting point.

To develop a design brief from either of these starting points you will need to think about the possible needs. This can be done by either using a mind map or making a list of your personal thoughts like the one in Figure 1.1.

To ensure that you have considered every possibility linked to your starting point, conduct research to help you find out actual user needs. This can be achieved most effectively through the use of a **questionnaire**. Remember to consider questions which analyse the following

- Jeans could be redesigned to make into a skirt or shorts – shortening of legs, restructuring leg side seams to stitch together as a skirt

- Time available, resources and cost considered – revamping and reuse would reduce demand on time and cost

- Creativity and design opportunities linked to jeans as a starting point

- Environmental, social and moral aspects linked to the starting point – recycling and reuse of an existing product are an advantage. Trends/morals in society are favourable towards environmentally friendly and sustainable textiles

- Fabric could be reused to develop a new product, e.g. hat, soft furnishing item, toy, piece of jewellery or a bag

- Traditional techniques used to recycle fabric could be explored, e.g. patchwork, quilting, appliqué and weaving

- Target groups considered – teenager, children (8–12 years), young professionals

- Disassembled parts could be re-sewn and integrated into a new product, for example a jacket with denim sleeves made from the leg patterns and pockets as a decorative feature

Figure 1.1 Analysing a starting point to develop a brief

headings: Who? What? How? and Where? (See Table 1.1.)

Once you have formalised your thoughts about your starting point and the user needs from your questionnaire results, a design brief can be written to summarise your aims. For example:

Starting point: old pair of jeans

Design brief: 'I shall design and make a textile garment for a teenage girl, made from a pair of recycled denim jeans.'

WHO?	WHAT?	HOW?	WHERE?
Who are the target age group? Who will use and buy the product?	What does the user want the product to do? What materials, styles, decorative features are preferred?	How will the product be used? (anthropometrics) How much will it cost? How will it be made?	Where will the product be used?

Table 1.1

KEY POINT

- A design brief is a written statement that identifies a problem to be solved. The design brief is used to encourage thinking of all aspects of a problem before attempting a solution.

ACTIVITY

Analyse and write a design brief for the theme 'Twentieth-century Inspiration'.

Analysis of the brief

Once you have written your design brief you need to consider what areas of research will be most helpful and relevant to you, before you start to formulate a design specification and design your ideas. Controlled Assessment Unit A571 in particular will expect you to be able to:

- show how existing textile products reflect and influence culture and society
- identify and compare how existing textile products can improve lifestyle and choice.

Identifying and comparing existing products

Designing a questionnaire to gather primary research about the lifestyle of and choices made by your user group is a good starting point to gather your ideas. Researching into existing products will help you to find out the types of textile products a consumer needs and will buy. It is useful to form a profile about your user group before you develop a specification, looking at lifestyle, buying habits, age range, gender, occupation, culture and profession. Other important factors to consider when evaluating products are:

- **The performance factor** – how does the user want the product or the fabric it is made from to perform?
- **The cultural factor** – how does a custom or a period in history influence textiles within the community today? For example,

Figure 1.2 Existing products are good resources for generating ideas

Figure 1.3 A student mood board looking at historical influences on children's wear

trends in 1970s fashion have made a comeback in 2008.

- **The social factor** – what effect does the influence of different religions, beliefs, places, pressures or desires have on an individual within society and how does this impact upon their needs?

- **The environmental and moral factor** – how do 'value issues' have an impact on an individual's lifestyle and on the products they use? For example, is it morally acceptable to advertise glamorous clothing in countries where people are living in poverty? How do economic, political, ethical, moral and environmental issues impact on the consumer? Are designers aiming at consumers who want a design-led product that is produced without hurting the environment, without employing sweatshop labour and is part of an ethical supply chain? The trend is catching on: in Britain chains such as Marks & Spencer, Debenhams, Sainsbury's, Tesco and Top Shop are selling many textiles that achieve these standards.

- **The aesthetic factor** – how do personal taste and judgement dictate the colours, patterns and shapes of products available on the market, and how are products developed to suit individual preferences?

KEY POINT

- Remember to ask yourself: Is the product needed? Whose needs were identified to produce it? Who benefits from the manufacture of the product? How will they benefit?

EXAMINER'S TIPS

- Visit local retail outlets and make notes on what is available for your user group related to your design brief.
- Consider historical links and themes. Look at the work of designers/fibre artists.
- Look at magazines and mail-order catalogues.
- Research internet sites for national information related to your theme.
- Photograph existing products already owned by your user group.
- Produce an in-depth product analysis and compare the differences between product types.
- Disassemble a product to understand its manufacture and the relationship between the materials used.

Remember to analyse all your findings in detail and explain which aspects of your research you will develop further into your specification and design ideas.

1.2 PRODUCT ANALYSIS

When explaining how a product reflects and influences culture and society and improves lifestyle and choice, it is important to scrutinise a product in the same way the designer or the manufacturer has. This can be done through a process known as **product analysis.** Find out how to analyse and evaluate a product by answering the following questions:

- **Methods of construction:**
  How has the product been stitched together? What type of seam has been used? How has the product been finished?

- **Legal requirements:**
  What legal requirements does the product need to meet?

- **Moral and environmental issues:**
  What impact could using these resources have on people or the environment?
  What happens to any waste produced during manufacture?
  What skills are needed to manufacture the product?
  What are the working conditions in the place of manufacture like?
  How easily can the product be recycled?
  Will the product have an impact on the environment?

- **Fibres, fabrics and components:**
  What fabrics and components have been used and why?
  Where do the materials and components come from? Will they run out?

What other resources have been used to manufacture the product?

- **Function:**
  How is the product used?
  Will the product have an impact on people's lives? Explain how.
  How long will it last? How will it be disposed of?
  What factors may limit or lengthen the product life?

- **Size and ergonomics:**
  What size and measurement is the product?
  How has it been designed so that it is easy/comfortable to use?

- **Design:**
  Is there a choice of designs, colour and shape? List them.
  How was the design developed/produced?
  What influences do you think had an impact on the design?

- **Your response to the product:**
  Do you want to touch or use it?
  Who is the product intended for? Why would they buy it?
  Would you like to own it? Why? What would this product say about you?
  Would this product enhance your lifestyle? Explain how.

A product analysis can be completed without 'taking apart' (disassembly) a product. Use digital photography to show your analysis of different sections of the item, for example seams, labels, fastenings, fabric construction, components used, surface decoration, etc.

ACTIVITY

1. Select two textile items, one made from a natural fibre and one produced from a manufactured fibre. Compare the products to show how each one impacts on environmental and moral issues.

2. List the advantages of evaluating existing products using product analysis.

EXAMINER'S TIPS

Data that you collect as research can be presented using:

- tables/charts/graphs – Excel ®
- digital photographs
- mind maps and annotated diagrams
- sketches and information boards.

1.3 ANALYSIS OF RESEARCH – THE DESIGN SPECIFICATION

Having devised a brief, researched and analysed your client profile, questionnaire results and existing products, you can then formalise a design specification. The design specification is a set of criteria that create the guidelines from which you will work to produce a functional product fit for purpose. Controlled Assessment Unit A571 and A573 both require you to produce a detailed specification.

Design specifications are produced by listing the important design criteria points and then fully justifying them. The following guideline points will help you create a detailed specification:

- **Time scale** of production for each controlled assessment unit – 20 hours for each unit. If you are following a short course you will only complete one controlled assessment unit (20 hours).

- **Function of the product** – what is the purpose of the product?

- **Performance** – how and where is the product meant to be used? What will it need to do? You could add a digital photograph of the problem to be solved here to help you explain the need.

- **Materials and their properties** – which materials will be suitable to use? What performance characteristics are preferred?

- **Appearance/aesthetics** – how will colour, texture, pattern, shape, etc. contribute to the visual appeal of the product?

- **Theme** – how does the product reflect the theme?

- **Anthropometrics** – what information about the user in terms of height, width, weight, reach, grip, movement, etc. needs to be considered?

- **Environmental and moral issues** – what legal and environmental requirements need to be adhered to? Can materials be recycled? Impact on people considered.

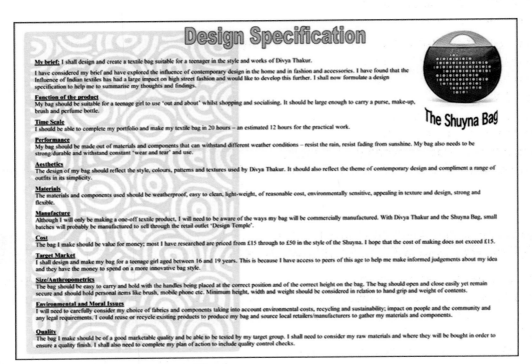

Figure 1.4 An example of a design specification

- **Target market** – which user group are you designing your product for?
- **Cost** – how much will the product cost to produce?
- **Health and safety** – what health and safety regulations are important to consider?
- **Quality** – how will you produce a quality product? What control points will you need to consider?
- **Size** – are there any specific measurements that are important to the function of the design?

Your design specification will help you to make decisions when you are:

- designing ideas for your final product
- developing and communicating ideas
- investigating and testing your ideas against suitable materials, techniques, components and production methods
- completing a critical evaluation of your product and the methods you have used.

EXAMINER'S TIPS

- Do not be afraid to use diagrams and digital photography in your specification to help you to justify your criteria points in a more creative way!
- Use subheadings to introduce each criterion point. For example:
  Sustainability
  The fabrics and components used will need to be carefully selected to ensure that they are environmentally friendly to produce, originate from a sustainable resource and/or can be recycled at the end of use.
- Specification points should always be fully justified.

ACTIVITY

Write a design specification for this portfolio outline: 'Soft activity toys challenge toddlers and encourage their development through creative play.'

1.4 PRINCIPLES OF GOOD DESIGN

What is meant by the term 'design'? There are a lot of definitions available to us, but in the context of your work within textile technology, design can be defined as being experiences that are evaluated and analysed by the user to create successful solutions that solve real problems. As a learner you will need to be able to:

- Identify associations linking principles of good design with technological knowledge.

What are the principles of good design?

The principles of good design are to make things look and feel good so that the consumer wants to use what has been designed, and also to ensure that the consumer can intuitively understand how to use the product and understand its purpose.

- Good design is innovative.

- Good design enhances the usefulness of a product.
- Good design is flexible: it accommodates a wide range of individual preferences, e.g. left-handed or right-handed scissors.
- Good design is aesthetic.
- Good design displays the logical structure of a product.
- Good design is unobtrusive.
- Good design is enduring – stands the test of time.
- Good design is consistent – correct to the details.
- Good design is ecologically conscious.
- Good design is simple and intuitive.

Design can often be taken for granted. A product works so well that the creative effort involved in making it has been forgotten: for example the sewing needle, so simple in its design it is obvious, yet at some point in history it was not! It is important to remember that every design has behind it a designer or designers who have tried to make the world a better place by solving a problem. Knowing and understanding your subject knowledge well is the first step to creating a quality product which displays good design features. Being able to use technology effectively in the design process will help you to achieve this.

1.5 DESIGN INSPIRATION AND INFLUENCES

This is the starting point for any design work. You can use a variety of methods to present your ideas and inspiration, including words and images. Design inspiration involves the

Figure 1.5 A student's piece of work inspired by cultural influence

collection of colour, texture, shape and pattern and is influenced by:

- the focus area – the theme or product you have selected to study

- justification of creative and/or innovative design

- the relationship to historical/cultural or contemporary factors – past designs, artistic and cultural influences.

CASE STUDY: ALEX RUSSELL CREATIVE SERVICES 1

Alex Russell specialises in printed textile design, surface pattern, graphics, illustration and other creative services for fashion, interiors and home furnishing. He mainly produces contemporary, innovative designs and artwork which are created using a powerful CAD (computer-aided design) system, traditional (hand) media or a combination of the two. Entire collections can be created, transferred into different colourways, illustrated and presented on technical specification sheets. He is also able to transfer designs into a repeat format or single colourways.

What inspires Alex when he designs depends on the project he is working on; in a lot of cases, the client will have a theme or idea that the designs have to fit into. This challenge is something Alex thrives on, asserting, 'Design is all about answering the brief (and answering on time)'. He looks for new ideas all the time, and more often than not his inspiration comes from outside fashion and textiles and more from illustration or graphic design. He finds it difficult to describe what inspires him the most, but it tends to be either things that look like nothing he has ever seen before or the complete opposite – trying to turn something very familiar into something new!

Alex believes that if you're serious about being a good designer, the three most important principles are; good drawing, colour and being able to

answer the brief. You need to be able to communicate your ideas as images and be able to pull visual information out of your inspiration sources, and drawing is central to both of these. Anyone can learn to draw – it just takes a lot of time and effort, but you have to practise all the time to become good, and keep practising to stay good. Colour is also very important. It's a really powerful method of creating a look and, as a professional, you have to be able to work with colours you might not choose yourself. There's a saying in textiles that you can sell a bad design if the colour is good, but you cannot sell a good design with bad colour; there's a lot of truth in that. Finally, it's always very important that the designs are right for what the client wants. When Alex works directly with a company, the brief is normally very specific and he has to answer all parts of it – that's what they pay designers to do.

Alex finds it very difficult to pick the design he would say was his most creative. He does have

Figure 1.6 An example of Alex Russell's work showing surface pattern designs for fashion and furnishing items using hand drawing and digital imagery

designs that stand out, either because he got the hang of a new technique as illustrated in Figure 1.6, or because the design was a surprise in some way, as shown in Figure 1.7.

Figure 1.7

The designs he creates normally end up being printed on to garments or home furnishings. Alex designs for all market levels from supermarket and high street to designer and couture. He is also able to design ideas for all end uses and users including men, women, children, sport and

swimwear, underwear and accessories, wallpaper, soft furnishings, floor coverings and carpets, plus fabric trends and prediction, inspiration and design direction. Some of the clients he has created designs for include Speedo, Osh Kosh (children's wear), Tommy Hilfiger, Etam and Levi Strauss.

Figure 1.8

1.6 DIGITAL MEDIA AND NEW TECHNOLOGIES

Digital technology is now a large part of everyday life and is a very important aspect of design within fashion and textiles. As a learner you will need to show that you:

- Have used ICT/CAD/CAM to support your design development
- Can apply appropriate knowledge of digital media and new technologies to your work.

The use of digital technologies like CAD/CAM, photography, video, image manipulation and analysis software, to collect and gather research and information data, is vital to the success of both of your controlled assessment units.

Figure 1.9 Gareth Pugh's Elumin8 jacket which features a geometric diamond pattern and strong angular shapes glowing with luminescent white light

Textiles have moved forward tremendously within the field of advanced technologies, from textiles that double as an electric circuit where bar codes and photographs can be woven into the fabric, to the electric plaid – a programmable handwoven textile that changes colour when an electric current is introduced. The most spectacular innovation is the development of flexible light that can be cut, folded and twisted, which was worked into an electroluminescent jacket designed by Gareth Pugh, unveiled at the London Fashion Week in 2005.

The use of digital ink-jet printers and image-based digital Jacquard weaving processes have helped to digitally enhance our textile products. In the classroom, printing from the computer on to transfer paper, which can then be ironed to transfer the image on to fabric is a simpler form of digital printing. Designs can also be transferred directly on to the fabric from the computer, using special printing inks, which is an ideal way to produce prototype designs. Textile designers like Kathy Schicker, Cathy Treadaway, Joan Truckenbrod and J.R. Campbell are all experts in the area of digital printing and manipulation on to fabric.

EXAMINER'S TIPS

When presenting your research and design work, think about using:

- digital photographs or video clips showing the product being manufactured and in use
- 2-D and 3-D modelling of design ideas using CAD software to help with image manipulation and editing.

Chapter 4 looks at these applications in more detail.

CASE STUDY: ALEX RUSSELL CREATIVE SERVICES 2

Alex first started working with textiles when he was 13 years old and was introduced to screen printing through his art lessons. By the time he reached A Level, he had started to experiment with computer graphics and decided to do a degree in printed textiles at the newly named Manchester School of Art. All his work at the time was either for interior textile or textile art and was hand-printed or painted. After Alex graduated, he worked as a freelance designer and lecturer, where he had access to digital technology, including Photoshop®, which remains the software he uses the most in his work today.

All Alex's designs start with some kind of drawing. Even though they are almost all digital by the time they are finished, everything begins with either pencil or paint and paper. Alex either scans these drawings straight into the computer as shown in Figure 1.10 or uses his sketches to work directly in either Photoshop® (Figure 1.11) or Illustrator® (Figure 1.12). Alex also works with his own photos (Figure 1.13) and on occasion he mixes all three (Figure 1.14).

Although he uses printouts to help sell his work, what the client actually uses is the digital image

Figure 1.10

Figure 1.11

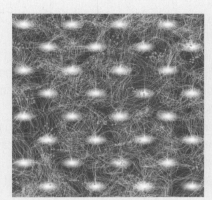

Figure 1.12

Figure 1.13

Figure 1.14

file. Adobe® software is industry standard and it does not matter if the work is completed using a Mac or a PC (although a lot of designers prefer Macs, this isn't always the case in textiles). Photoshop® is probably the most common software, but more and more clients prefer Illustrator®.

In 2001 Alex established his own business and started to build up a collection of print and pattern designs for fashion. He has sold his ideas and designs to clients all over the world, and still loves it when he sees someone wearing one of his designs. He also does fashion illustration and trend and inspiration books for fashion.

ACTIVITY

1. Explain why you think Alex likes to combine hand illustration with digital imagery in his work.

2. Produce a report outlining the advantages and disadvantages of digital imagery in the presentation of design and research information.

CASE STUDY: BERNINA – 'MYLABEL' 3-D FASHION PATTERN SOFTWARE

Figure 1.15 MyLabel logo

This virtual fashion technology developed by Siemens PLM Software, OptiTex and Bernina has already had an impact on home sewing and pattern making in the classroom. Bernina 'MyLabel' allows the user to customise fabric choices, embellishment and stitch choices and fit a garment idea to a 3-D mannequin that conforms to personal body measurements. The design pattern can be printed ready to sew with step-by-step instructions included. 3-D animation allows the user to view all pattern pieces with both the selected fabric and embellishments highlighted.

The software guides the user through the entire process from concept to completion covering the following stages:

- The user inputs their personal measurements and the 3-D dress form or mannequin 'morphs' into a body-size reflection of the user.
- A Style Consultant facility is available to help style choice (20 built-in styles). An onscreen information tab can be used to tell the user exactly how and where to measure.
- A large fabric library is available for the user to select a suitable fabric *or* the user can scan in a fabric swatch to view on the garment.
- A wide choice of buttons, embroidery and machine stitches is available.

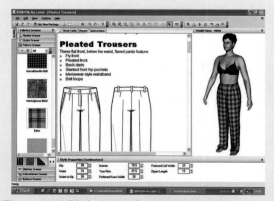

Figure 1.16 MyLabel software gives you the opportunity to create a varied wardrobe that reflects you: your taste, your interests and your body. You are able to visualise the result even before cutting into your fabric!

- After the garment has been simulated, a tension indicator can be used to see where the fit is loose and closer to the body. This can be changed to make the fit better for the wearer by using the style properties of the software.
- Once the garment is complete, the user can save, print or email the customised 3-D model to be stored and used at a later date.

- The pattern can be printed and then taped together or the user can take the file to a copy centre and have it printed out on a wide-scale printer.
- Step-by-step instructions are also available to be printed out for each pattern.

1.7 DESIGNING THE PRODUCT

LEARNING OUTCOMES

In this section you will learn about:

- generating and recording a range of innovative design solutions for a specific task/user need
- evaluating and modifying ideas, considering creativity and sustainability
- developing and modelling design proposals and justifying chosen ideas in relation to designs and materials
- using a range of skills to communicate ideas using graphic techniques, ICT and digital technologies
- the purpose of prototyping when designing, using a variety of materials
- the principles of anthropometrics when designing.

The word 'design' means to create or execute in an artistic or highly skilled manner – in the way Chanel designed the famous 'little black dress'. Similarly, your controlled assessment portfolio requires you to produce 'creative and original' ideas based around a product or theme. You will be expected to develop and present these ideas using appropriate strategies. It is therefore important to understand the nature of design work and the ways in which it can be presented.

How can designs be innovative and creative?

Designing is the combining of known elements in new and exciting ways in order to create innovative and fresh ideas. It is useful to remember that the main elements of design are silhouette, line and texture, and it is good practice to research into the work of other designers/manufacturers and spot trends in the market, to give you ideas on which to base your designs.

How to start:

- Select a 'starting point' that you are interested in for your controlled assessment portfolio; this will help you to generate appropriate and creative ideas.
- Understanding how materials work together and making the right fabric choice in relation to function and aesthetics will also help you to design creatively.
- Do not be afraid to let a fabric stimulate your design ideas, particularly if you are studying a theme linked to recycling.
- Identify a user group that you have access to and can involve in the design process.
- Gather research from classic textile ideas, different cultures, historical backgrounds, social influences, lifestyle themes and other sources of inspiration; use these ideas to form a mood board or worksheet(s) that will develop into your designs.
- Experiment with ways to express your design ideas, try collage, drawing, photography and CAD manipulation.

Mood boards/storyboards

In industry, the designer uses a mood board to present a range of ideas through artwork alongside a prototype idea. The designer collates magazine pictures, articles, photographs, fabrics, illustrations and written information explaining about the theme, colours and the target market. Adding samples of new yarns and creating swatches using different techniques gives the designer and the client an idea of how the yarn feels, looks and handles. The mood board can form

the core of your portfolio if it is presented well and justified in its use.

Figure 1.17 Storyboards are used by designers to present a solution to a design brief

Initial ideas

Initial ideas or design sketches are a lot of basic ideas completed quickly, usually using a pencil. In industry these can also be referred to as roughs or design developments. These sketches can be explored, discarded or developed further and are useful for experimenting with different styles, patterns and colours. A digital camera can be used at this stage for capturing variations of designs, which can then be stored and developed further using CAD.

Working drawings

Working drawings or presentation sketches show your completed design ideas. These are usually coloured sketches containing details about the materials to be used, shape, size, ideas for trim details, fastening ideas, construction methods and decorative techniques. Working drawings show how your design ideas meet the needs of the user and your product specification. Important details need to be highlighted through clear annotation.

The final design

You need to consider carefully which design idea covers all your specification criteria. Your final decision should be thoroughly reasoned, using clear annotation and links to your specification. This can be achieved in the following ways:

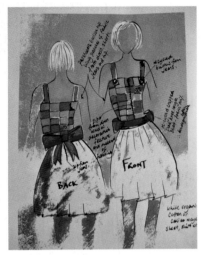

Figure 1.18 An example of how a textile student has formulated a design idea using sketching and modelling techniques

- detailed drawings of the final product (front and back views)
- the use of a chart, graph or star profile to show your links to your specification
- clear and precise written information or notes explaining your reasons for choice
- modelling important decorative and construction details or prototyping parts of a product, using different materials to test suitability
- full details included about fastenings, fabrics, assembly ideas and seams.

Anthropometrics and design

When making decisions about the choice of your final product you will be evaluating its fitness for purpose and how it can be better designed to suit the needs of the user. Anthropometrics is the study of people and their measurements, for example the size of the head, the size of the hands, length of the leg, etc.

The designer needs to collect anthropometric data to check product suitability. Accurate

EXAMINER'S TIPS

When sketching and drawing:

- Produce a wide range of sketches initially and then refine your ideas to between six and eight ideas.
- Make sure your designs are realistic and sustainable.
- Draw different viewpoints of your design ideas, e.g. front, side and back, to show design and construction features. Include details such as darts, gathers and pockets.
- Annotate important details and add swatches and samples of fabrics, components and worked techniques.
- Include measurements and dimensions on your final drawing.
- Present your ideas using a range of techniques and skills, e.g. modelling, ICT, free illustration, etc.

data related to the height, weight, limb and body-segment size of an individual are needed to design items like clothing and furniture, to ensure a quality product is produced which is suitable for its intended purpose.

KEY POINTS

- Use anthropometric data collected from research to base your design ideas around and to test the product with your target group.
- Designing is about communication and the conveying of ideas. It is important therefore that your designs are clear and accurate.

ACTIVITY

1. Produce a mood board/storyboard using an aspect of ICT, based around one of the following themes; camouflage, extreme sports, denim or historical origin.

2. Take a starting point from your mood board and explore your design ideas. Present two working drawings and give detailed reasons for your choice.

1.8 PRODUCT PLANNING

LEARNING OUTCOMES

In this section you will learn about:

- producing a detailed plan for making a textile product
- choosing and preparing materials economically, considering cost, sustainability, the environment and moral and cultural issues
- planning work to make the best use of materials, components, equipment and resources, including time and energy
- strategies used to help overcome problems that arise during production.

Thorough planning when making your final product and presenting your controlled assessment portfolios is very important.

Time management

When planning the production of your textile item you will need to build in time to complete the following:

- trialling and testing of materials and pre-manufactured components
- trialling and testing construction processes and decorative techniques
- how you will organise the time you have available to complete the task
- what tools and equipment are available to you and most suitable for the product
- preparation of a toile in order to check control system

- control checks and possible problem areas
- supportive photographic or video evidence of each key stage of production
- health and safety
- environmental, moral, social and economic implications.

A work schedule or plan for making should include information or decisions relating to all of the points listed above. You will have to decide what needs to be done and in what order you are going to do it, recording your successes and failures in order to 'review and reflect', to ensure successful completion within your allocated deadline.

Planning your resources

For your controlled assessment portfolios you will be expected to show that you can plan and organise activities within your making, to ensure that you select appropriate materials and tools (hand and machine) to successfully make your textile product. You will also be required to record key stages in the making of your product through written and photographic evidence. It is vital therefore that you carefully check that all the resources needed are available.

- Materials must be carefully chosen to fit with the product specification. Fabrics and components which may look aesthetically suited to the purpose of the product may prove not to be when in use.
- Materials need to be prepared economically, allowing for wastage.

Once you have identified the resources you will need to make your product, you can start to plan an order of work.

Work schedule/flow chart

You will find that the best way to present your plan of action is through a flow chart. This format helps you to plot where you need to make a decision about a process or technique and where you need to apply a quality-control check. This helps you to monitor your work in progress, ensuring that you produce a quality textile item which is fit for purpose.

Resources and health and safety issues can also be added to this chart, alongside a 'time tracking system' which will help you to show how long it will take you to make the product from start to finish.

A deadline for completion will be set for your controlled assessment portfolios. The total time allowed is 20 hours for each controlled assessment portfolio, with the planning and making section accounting for nearly half of the total marks. This suggests that approximately 8–10 hours should be allowed for planning and making of the final textile product.

Remember your time plan will need to take into account the following points:

- start and finish dates/timings
- type and quantity of fabrics needed, considering cost, sustainability, environmental, moral and cultural issues
- type and quantity of manufactured components, e.g. thread, zip, elastic, etc.
- preparation and clearing-away time
- quality-control checks
- any unforeseen problems
- availability of tools and equipment in the classroom.

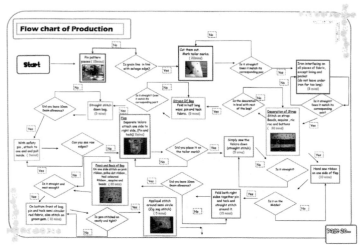

Figure 1.19 An example of a student's flowchart to include photographic evidence of making and timings

Your controlled assessment portfolios can be created as a team task. It is important to remember, however, that each member of the team *must* identify which part of the portfolio they have completed for marks to be awarded consistently and accurately.

In industry, product planning and manufacturing can be broken down into stages to include:

- preparation or sub-assembly, where items are bundled ready for sewing or small components may be attached to specific sections of a garment
- processing
- assembly and finishing
- packaging.

At the end of the process the final textile item is checked for any defects before despatch. All the work plans, construction details, time schedules are completed using CAD programmes. This helps to save time and money and ensures accuracy and quality.

Product costing

It is important to consider cost when planning the making of your product. Costs can be divided into two main areas:

- cost of conformance – efficient quality assurance is in place and no faulty goods produced.
- cost of non-conformance – the cost resulting from faults and errors in the manufacturing process, e.g. cutting out the pattern pieces incorrectly.

Remember when you are planning the making of your product to consider the cost of non-conformance. This needs to be allowed for in the calculation of your material amount. In industry costs are calculated using direct costs, which are the actual cost for making the product, and indirect costs, which cover all other costs that are incurred in the completion of the product to a marketable standard, e.g. the running of machinery, staff wages, advertising, etc. This is why a bespoke garment is so expensive to make, compared to a similar product manufactured using batch production where the costs can be shared.

KEY POINTS

- When selecting resources, environmental considerations are important. Think about the implications of industrial manufacture, e.g. the making process and finishes applied to fabrics, washing and cleaning, recycling and sustainability implications. The Practical Action website (www.sda-uk.org) and *The Eco-design Handbook* by Alastair Fuad-Luke (Thames and Hudson) are useful resources.
- Consider health and safety issues when making your product, both in the classroom and in industry.

ACTIVITY

1. Complete a work schedule/flowchart for a product of your choice. Remember to include photographic evidence, timings and quality check points with your sequenced plan.

2. Identify the problem areas that may cause deadlines to be missed.

CHAPTER 2

MATERIALS

By the end of this chapter you should have developed a knowledge and understanding of:

- The various fibres used to make fabrics, and their performance characteristics
- The conversion of fibres into yarn and fabric
- The mixing and blending of fibres and yarns
- The different finishing processes that can be applied to fibres and fabrics
- Methods of colouring fabrics
- Modern materials, including 'smart' materials and new developments in fabrics
- Pre-manufactured components

2.1 FIBRES

In this section you will learn about:

- the various fibres used to make fabrics.

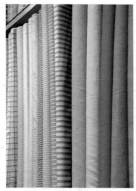

Figure 2.1 A selection of fabrics

There is a huge range of fabrics on the market. Each fabric is unique in the way it looks, known as aesthetics, and the way it performs and the qualities it has. Some fabrics are designed to be warm to wear, others to keep you cool. Some are designed to be hard-wearing, while others are waterproof. These qualities are known as 'performance characteristics'. When making a textile product it is important to choose a fabric that not only looks good but also has the right performance characteristics.

EXAMINER'S TIP

In the examination you will need to be able to name the fibres used to make fabrics and identify the performance characteristics of each. You will need to be able to suggest specific fibres and fabrics for particular textile products and give reasons for your choices.

Performance characteristics of a fabric

The performance characteristics of a fabric are determined by a number of factors:

- the fibre used to make it
- the way the fibres are made into yarn
- the way the yarn is made into fabric
- any special finishes applied to the fabric.

The origin and structure of fibres

A fibre is a fine, hair-like structure. The first fibres used to make fabric were natural fibres – they came from plants and animals. In the 1940s and 1950s, fibres were made for the first time from chemicals, known as synthetic or manufactured fibres. A third group of fibres, known as regenerated fibres, are made from a natural starting point, which is then treated with chemicals to make a fibre. Modern technological developments have resulted in the production of 'smart and modern' fabrics with a whole new set of performance characteristics. The origin and structure of the fibre determine its performance characteristics.

Natural fibres

These fibres come from plants and animals and are often considered to be environmentally friendly. They are renewable resources and can be produced organically, but you should not assume that they are all produced in this way.

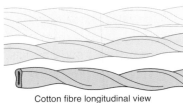

Cotton fibre longitudinal view

Cross section

Mature   Immature

Figure 2.2 A cotton plant and cotton fibres

Cotton fibres

These fibres come from the cotton boll – the seed pod of the plant. The fibres grow around the seeds in a hard casing that looks like a conker. The seed pod eventually bursts, making the fibres visible. They are quite short fibres, known as staple fibres, and are made from cellulose.

Linen fibre

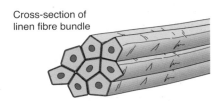

Cross-section of linen fibre bundle

Linen fibre (flax) cross section

Figure 2.3 A flax plant and linen fibres

Flax or linen fibres

These fibres come from the stem of the flax plant and look rather like bamboo. They are longer than cotton fibres but have similar performance characteristics as they are also made from cellulose.

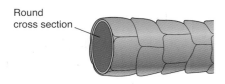

Figure 2.4 Sheep and wool fibres

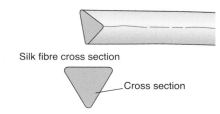

Figure 2.5 A silk worm and a cocoon with a silk fibre

Wool fibres

Wool fibres are animal hairs. They are usually collected from sheep, but can also come from goats, llamas, camels and angora rabbits. Wool fibres are like human hair and are made from protein. Wool fibres are classified as staple fibres, but the longer fibres make better quality fabric than the very short ones.

Wool fibres have scales on their surface. These scales can trap air, making fabrics made from wool fibres warm to wear, but they can also become tangled if woollen fabrics are not washed carefully, causing the fabric to become hard and shrink.

Silk fibres

Silk fibres come from the cocoon of a silk worm. A silk worm is actually a caterpillar that spins a cocoon around itself before it turns into a moth. The cocoons are dropped into boiling water to kill the caterpillar and loosen the 'glue' used to stick the silk fibre together to form the cocoon. The cocoon can then be unravelled, giving a long silk fibre that can be a thousand metres long. Silk is the only natural continuous filament fibre and is made of protein.

If the silk worm is allowed to emerge from the cocoon, it 'eats' its way out of one end and breaks the long continuous fibre into thousands of short lengths. These short fibres can be spun into silk yarn, but this will not have the same lustrous qualities of the long unbroken fibre, nor will it be as strong.

▌ Regenerated fibres

Regenerated fibres are made from a natural starting point, such as wood pulp or very short cotton fibres, called linters, which are too short to spin into yarn. These provide cellulose, which is treated with chemicals to produce a thick, sticky substance rather like treacle. This liquid is forced through a machine called a spinneret, which has a series of holes in it rather like a shower head. The thick liquid comes out in strands and is solidified either by passing through a chemical bath (known as wet spinning) or by cooling in the air (known as dry spinning). The size and shape of the holes in the spinneret can be changed to create fibres with different characteristics. The solidified fibres can be used in long lengths – filaments – or cut into shorter lengths – staple fibres.

Examples of regenerated fibres are viscose, acetate, triacetate, modal and lyocell.

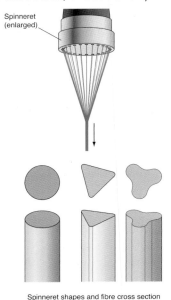

Spinneret shapes and fibre cross section

Figure 2.6 A spinneret and examples of fibre cross sections produced

Viscose fibres

This is the most commonly used regenerated fibre. It is made from pine, beech or eucalyptus wood, dissolved in a chemical, then passed through a spinneret. The fibres produced are solidified in a chemical bath. The fibres are usually cut into staple lengths before being spun to make a yarn, as this improves the performance characteristics for the final fabric. Viscose is a cheap fibre to produce and feels like cotton.

▌ Manufactured fibres

These fibres are made purely from chemicals, usually from the coal and oil industries. They are made from non-renewable resources and are not biodegradable and are therefore not considered to be environmentally friendly.

During manufacture, the fibres can be adapted to meet a range of different needs and engineered to have the qualities needed for a particular product. The thermoplastic qualities of these fibres make them particularly versatile. They are the most innovative of all fibres and are continually being developed for new uses.

Figure 2.7 Oil is used to make synthetic fibres

Polyester fibres

This is the most important and most widely used synthetic fibre. It is inexpensive to manufacture, and is usually cut into staple fibres to improve its performance characteristics.

A recent development has been to make polyester fibres from plastic bottles. This fabric has a trade name of Polartec® and is often produced as a fleece fabric. It takes 25 plastic bottles to make one Polartec® jumper.

Figure 2.8 Plastic bottles are used to make Polartec®

Polyamide fibres

Polyamide fibres are often referred to as nylon and were the first synthetic fibres to be developed, initially intended to imitate silk fibres. They are produced in filament form but as with other synthetic fibres they are often cut into staple lengths to improve their performance characteristics. As the fibres come out of the spinneret, they are stretched slightly to produce a fine, strong fibre.

Acrylic fibres

These fibres handle like wool fibres, but are easier to wash. They are inexpensive to produce and are often used to make products traditionally made from wool fibres such as blankets, knitting yarns, fake fur and fleece fabrics.

Elastane fibres

The most well-known elastane fibre is Lycra® and its most well-known performance characteristic is its ability to stretch. Elastane fibres can stretch by up to 500 per cent and return to their original length. They are produced as a filament yarn and are used in a 'raw' state in sheer hosiery, foundation garments and medical supports.

Most elastane fibres are covered with another yarn before they are used in garments. This improves comfort and wear, as the fibre does not come into direct contact with the skin. Elastane fibres increase the amount of stretch in a fabric, providing some elasticity and improving crease and wrinkle resistance. They are fine, washable and easy to colour.

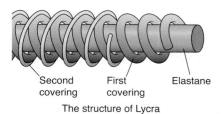

The structure of Lycra

Figure 2.9 An elastane yarn

2.2 PERFORMANCE CHARACTERISTICS OF FIBRES

In this section you will learn about:
- the performance characteristics of fibres.

The term 'performance characteristics' relates to the way fibres behave and the qualities they have. It is important to consider these when choosing a fabric for a textile product, but it is also important to bear in mind these characteristics will be influenced by the way the fibres are made into yarn and fabric and also if any special finishes have been applied.

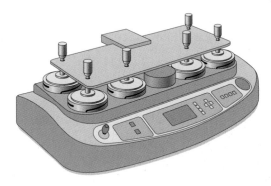

Figure 2.10 A Martindale abrasion testing machine

Abrasion resistance

This refers to the amount of rubbing, wear and tear a fibre or fabric can withstand. Fibres and fabrics used for furnishings, floor coverings, bags, bedding, school uniforms and workwear need to withstand a lot of wear and tear and therefore need to have good abrasion resistance. When textile items are washed, they suffer a certain amount of abrasion, or rubbing, so items that are frequently washed also need to have good abrasion resistance.

Strength

This refers to the amount of pulling a fibre or fabric can withstand. For example, the handles on a shopping bag will have to be able to hold the weight of the bag, which will increase as items are put into it.

Figure 2.11 Shopping bag handles need to be strong

Elasticity

All fibres and fabrics have a certain amount of elasticity, or 'give', not to be confused with the qualities of elastic. The more elasticity a fibre or fabric has, the less likely it is to crease, and the more comfortable it will be to wear.

Absorbency

Fabrics that have good absorbency are usually comfortable to wear, as they absorb perspiration, which then evaporates away from the fabric. Fibres and fabrics used for towels need to be absorbent if they are to function. Fibres and fabrics that hold moisture are also less likely to build up static electricity. Static electricity causes fabrics to cling to the body and also to attract and hold dirt particles, neither of which are desirable qualities. Fabrics that are absorbent take a long time to dry.

Figure 2.12 Towels need to be absorbent

Thermal conductivity

Fibres and fabrics that have good thermal conductivity are cool to wear. Heat is conducted away from the body by the fibres and fabric. Fibres and fabrics with poor thermal conductivity are warm to wear, as they trap heat next to the body.

Washability

The washability of fibres and fabrics depends on a number of other performance characteristics such as abrasion resistance, the ability to withstand heat, ability to withstand chemicals, crease resistance and how absorbent a fabric is. This is an important factor to consider when choosing a fabric for an item that will need to be washed frequently.

Figure 2.13 A fleece top needs to be warm to wear: the fabric used has poor thermal conductivity

Resistance to chemicals including acid, alkaline and bleach

Chemicals are regularly used when washing textile items. Washing agents and fabric softeners as well as stain removal agents are chemicals, and it is important to look at the labels on these products before using them on a fibre or fabric. Some fibres and fabrics are particularly sensitive to chemicals and need to be washed carefully.

Perspiration is a mild acid, as are some fruit juices and vinegar. Leaving these acids in contact with some fibres and fabrics for prolonged periods could cause permanent damage, so the item should be washed as soon as possible.

Bleach can be safely used on some fibres and fabrics, but it is advisable to test it on an inconspicuous area first before treating the whole item.

Figure 2.14 Detergents are chemicals

Flame resistance

Some fibres and fabrics catch fire and burn more easily that others. Fibres and fabrics that have a lot of air trapped in their structure will burn more easily. Some fibres and fabrics smoulder and melt rather than burn, but this can be just as dangerous. There are special

finishes that can be applied to fibres and fabrics to reduce their flammability.

Moth and mildew resistance

Moths can be a problem for protein fibres such as wool and silk because they lay their eggs in clothes that are stored. When the larvae hatch they use these natural fibres as a source of food, making holes in the fabric.

Mildew is a form of mould that flourishes where there is high humidity such as bathrooms and shower rooms as well as in humid countries. Fibres and fabrics with a high cellulose content such as cotton and linen are particularly at risk, and wool and silk fibres can also be attacked. Synthetic fibres are immune.

Thermoplasticity

Only synthetic fibres and fabrics have this performance characteristic. They can be heated up and set into a shape that they maintain when they cool down. This makes some finishes easy to apply, such as texturing and bulking of fibres and yarns and also means fabrics can have pleats permanently set in.

Figure 2.15 Children's nightwear must be flameproof

Figure 2.16 A pleated PE skirt needs to retain its pleats when it has been washed

ACTIVITY

Make a collection of different textile products. Photograph or draw them. Next to each illustration, list the performance characteristics needed by the product and suggest a fibre or fibres that could be used to make it. The table of performance characterstics shown in Figure 2.17 will help you with this.

Performance characteristics	Acetate	Acrylic	Cotton	Linen	Polymide (nylon)	Polyester	Silk	Triacetate	Viscose	Wool
Abrasion resistance	✓	✓	✓✓	✓✓✓	✓✓✓✓	✓✓✓✓	✓✓	✓	✓	✓✓
Absorbency	✓✓	✓	✓✓✓	✓✓✓	✓	✓	✓✓✓✓	✓✓	✓✓✓✓	✓✓✓✓
Crease resistance (stretch)	✓✓	✓	✓✓✓	✓✓✓	✓✓✓✓	✓✓✓✓	✓✓	✓✓✓	✓✓	✓✓✓✓
Flame resistance	✓	✓	✓	✓	✓✓	✓✓	✓✓✓✓	✓	✓	✓✓✓✓
Insulation	✓✓	✓✓✓	✓	✓	✓✓✓	✓✓✓	✓✓✓	✓✓	✓✓	✓✓✓✓
Moth resistance	✓✓✓✓	✓✓✓✓	✓✓✓✓	✓✓✓✓	✓✓✓✓	✓✓✓✓	✓	✓✓✓✓	✓✓✓✓	✓
Mildew resistance	✓✓✓✓	✓✓✓✓	✓	✓	✓✓✓✓	✓✓✓✓	✓	✓✓✓✓	✓	✓
Resistance to acids	✓✓✓	✓✓✓✓	✓✓	✓✓	✓	✓✓✓✓	✓	✓✓	✓✓	✓✓✓
Resistance to alkalis	✓	✓✓✓	✓✓✓✓	✓✓✓	✓✓✓	✓✓✓✓	✓	✓✓✓	✓✓✓	✓✓
Resistance to bleach	✓	✓✓✓	✓✓✓	✓✓✓	✓✓✓	✓✓✓✓	✓	✓	✓	✓
Resistance to build-up of static electricity	✓	✓	✓✓✓✓	✓✓✓✓	✓	✓	✓✓✓✓	✓	✓✓✓✓	✓✓✓✓
Tensile strength	✓	✓✓	✓✓✓	✓✓✓	✓✓✓✓	✓✓✓✓	✓✓✓✓	✓	✓✓	✓
Thermal conductivity	✓	✓	✓✓✓	✓✓✓✓	✓	✓	✓	✓	✓✓	✓
Thermoplasticity	✓✓✓✓	✓✓✓✓	n/a	n/a	✓✓✓	✓✓✓✓	n/a	✓✓✓	n/a	n/a

Figure 2.17 Table showing the performance characteristics of fibres

2.3 CONVERSION OF FIBRES INTO YARN AND FABRIC

In this section you will learn about:

- the various ways of making fibres into fabrics
- the performance characteristics of different fabrics
- the names of specific fabrics and their uses.

Fibres can be converted directly into fabric, but most are made into yarn first. The way the fibres are made into yarn, along with the fibres used, determines the performance characteristics of the yarn.

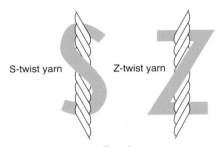

S-twist yarn   Z-twist yarn

Figure 2.18 S twist and Z twist

Spinning

Plain yarns

Spinning is the drawing out and twisting of fibres together to form a yarn. If the yarn is twisted a lot, most of the air between the fibres will be squeezed out and the fibres will be in close contact with each other. The yarn produced will be fine and strong, but will not be warm. If the fibres are loosely twisted, the air will remain trapped between the fibres, making the yarn warmer, but as the fibres are not in close contact, the yarn will be weaker.

The yarn can be spun clockwise, giving a Z twist, or anticlockwise, known as an S twist. These different twists reflect light in different ways, so interesting effects can be achieved when they are made into fabrics.

There are two different types of spinning systems, the worsted system and the woollen system. The worsted system uses longer fibres, which are thoroughly combed so that they lie parallel before they are twisted. This produces a smooth, regular, hard-wearing yarn that has a slight sheen to it as light is reflected off it.

The woollen system can be used for any fibre, not just wool, and the fibres used can be quite short. The fibres are arranged to lie roughly parallel before they are twisted, but are not combed as in the worsted system. This produces a coarse, hairy yarn, which is not as strong or shiny.

These yarns are called single yarns. Two or more single yarns can be twisted together to form a multi-ply yarn. If two single yarns are twisted together, it is known as a two-ply yarn, three yarns twisted together make a three-ply yarn, and so on. Up to twelve yarns can be plied together forming a strong yarn which can be even stronger if a Z twist is plied with an S twist.

Even multi-ply yarns can be twisted together to form a 'corded' yarn, which has a wide range of uses from embroidery threads to boating ropes. All of these yarns are 'plain' yarns, as they are even and regular along their length.

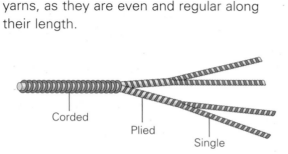

Figure 2.19 Multi-ply and corded yarns

Fancy yarns

These yarns are irregular, changing thickness and texture along their length. They may have bumps, knots and added fibres along their length, making them interesting but difficult to work with. They can be used to make fabrics either by weaving or knitting, but as they are not as strong as plain yarns the fabrics are not as strong either.

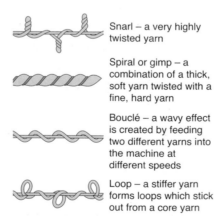

Snarl – a very highly twisted yarn

Spiral or gimp – a combination of a thick, soft yarn twisted with a fine, hard yarn

Bouclé – a wavy effect is created by feeding two different yarns into the machine at different speeds

Loop – a stiffer yarn forms loops which stick out from a core yarn

Figure 2.20 Fancy yarns

ACTIVITY

Make a collection of different types of yarns and study them using a hand lens or microscope. Describe their structure and suggest uses for each.

Weaving

Woven fabrics are made by interlacing two sets of yarns at right angles to each other. Weaving is done on a loom. One set of yarns, called warp yarns, is put on to the loom first. These yarns need to be strong to withstand the stresses and strains of the weaving process. A second set is woven in and out of this first set, travelling across the loom from right to left and back again. These yarns are called weft yarns and do not need to be as strong. They are carried across the fabric in a shuttle, a gripper, a jet of air or even a jet of water. At the edge of the fabric, where the weft wraps round the last warp yarn a selvedge is formed.

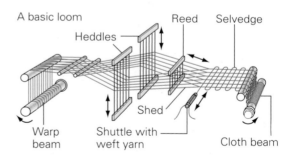

Figure 2.21 A loom

There are countless variations of woven fabrics, but they all have certain characteristics in common. The two sets of yarns are visible in the fabric, one set running up and down, the other set running from left to right.

The fabric will be strongest along the 'straight grain' of the fabric following the warp yarns.

The fabrics are generally firm and stable. They do not stretch much along the yarns, but they do stretch diagonally across the yarns, known as the bias of the fabric. When cut, the raw edges of the fabric 'fray' – yarns pull out of the fabric causing it to eventually disintegrate.

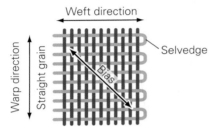

Figure 2.22 A woven fabric showing warp, weft, bias and selvedge

The characteristics of woven fabrics can be changed in a number of ways. The fibre used to make the yarns and how those yarns are made into the fabric play an important part. Plain and textured yarns give different effects when combined in different ways. The colour of the yarns can be varied to create checks or stripes in the fabric, and the thickness of the yarns can create ribbed effects in the fabric. The spacing between the yarns is very important. If the yarns are placed close together, a strong, firm fabric will be made but it will be expensive to produce. If the yarns are placed further apart, a cheaper fabric will be made, but it will not be as strong and will pull out of shape easily. However, as more air is trapped in this fabric it will be warmer to wear.

Varying the order in which the weft yarns interlace with the warp yarns changes the appearance and performance of the fabric. A special finish can be applied to the fabric to add any additional performance characteristics required.

Plain weave

Plain-weave fabrics have an even surface and look the same on both sides. This is the most basic weave and the cheapest to produce. The weft yarn passes over one warp yarn and then under the next, then over, then under, and so on. When the weft yarn travels back across the loom, it passes under the warp yarns it went over, and over the ones it went under on the previous row.

Plain-weave fabrics are smooth and even, so they are easy to print on. As the yarns interlace often, the fabric is less likely to crease, although it does not drape as well as some other weaves. Depending on the fibre and yarn used, it can be a hard-wearing weave. The spacing between the yarns will determine how thick or soft the fabric is.

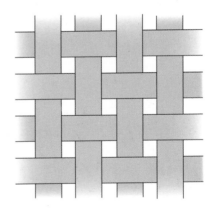

Figure 2.23 A plain weave

Examples of plain-weave fabrics are calico, poplin, muslin, lawn, shantung and rip-stop nylon.

Twill weave

Diagonal lines or 'wales' are visible on the surface of this fabric. The weft yarn passes over and under more than one warp yarn at a time. In the following row, the pattern moves along one yarn to create the diagonal pattern.

There are more variations possible with this hard-wearing weave, but it is more expensive to produce than a plain weave. The front and back of twill-weave fabrics look different, and as the surface is uneven it shows the dirt less. As the yarns interlace less often than in a plain weave, the fabric is less likely to wrinkle and the yarns can be packed closer together to give a firm fabric. Twill-weave fabrics are more likely to fray than plain-weave fabrics.

Examples of twill-weave fabrics are denim, drill, gaberdine and serge.

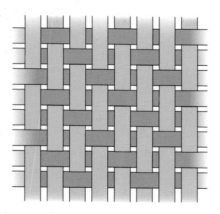

Figure 2.24 A twill weave

Satin weave

The right side of these fabrics is smooth and shiny, and the fabrics drape well. The weft yarns pass under at least four and sometimes as many as eight warp yarns before going over one. As the warp yarns lie on the surface in relatively long lengths, they 'snag' and catch easily, which means the fabric is easily spoiled. The fabrics also fray readily, as the yarns do not interlace much. The warp yarns are set close together and almost completely cover the weft yarns, making the fabric smoother, stronger, stiffer and more likely to wrinkle than a loosely woven fabric. This is an expensive weave as the number of yarns per centimetre is high compared to other weaves.

Examples of satin-weave fabrics are duchesse, satin, sateen and damask.

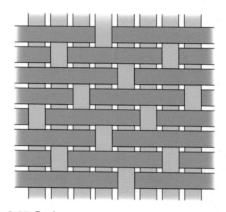

Figure 2.25 Satin weave

Jacquard weave

These weaves are very complex and are often made on a computer-controlled loom. Each warp yarn can be lifted individually and intricate designs and patterns can be created within the fabric. The fabrics made are very good quality and expensive.

Pile fabrics – loop and cut

These fabrics have threads or loops on the surface of the fabrics, made by an additional warp or weft yarn being woven in to the fabric. The main fabric is called the 'ground' fabric, and the additional yarns form a loop on the surface. The loop can be left as a loop, as in terry towelling, or cut, as in velvet.

Pile fabrics are thicker than most other fabrics and are therefore more hard-wearing. The surface of the fabric has a texture, so is more ornamental. However, care must be taken when working with pile fabrics, as they can seem to be a different colour depending on which way round they are. It is important to lay all pattern pieces on the fabric the same way round when making a product so that an even colour is achieved. Pile fabrics can be quite difficult to work with.

As air is trapped in the pile, these fabrics are warm to wear. In the case of terry towelling, the loops increase the surface area of the fabric, making it more absorbent and therefore particularly suitable for towels. The denser the ground fabric and the pile, the better the quality of the fabric. The ground fabric can be knitted as well as woven, and a cheaper way to make the fabric is to push tufts of yarn or fibres into the ground fabric. These fabrics may be cheaper, but the quality is not as good and the tufts of yarn or fibre fall out easily.

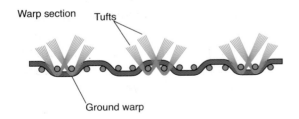

Figure 2.26 Corduroy

Examples of pile fabrics are terry towelling, velvet, velveteen, corduroy and fake fur.

ACTIVITY

1. Make a collection of woven fabrics. Cut a small sample from each and mount them on paper.

2. Next to each sample write down the weave used to make it, and if possible the fibre content too.

3. Find a picture of a product that would be made using that fabric and explain why it is suitable for that use.

Knitting

Knitted fabrics are made by creating loops in yarn and interlocking them. They are generally more stretchy than woven fabrics and do not fray when cut. As the loops in the fabric trap air, they are usually warmer to wear than woven fabrics. The loops are usually visible in the fabric, although fine machine-knitted fabrics have small loops, which may be difficult to see.

There are two types of knitting: warp knitting, where the loops are interlocked vertically, and weft knitting, where the loops are interlocked horizontally.

Warp knitting

In this fabric, the loops are linked from side to side, vertically up the fabric. They can be made on straight or circular knitting machines. Each needle on the machine is fed with its own yarn, so it takes a long time to set up a machine, but once ready up to 50 metres of 4-metre-wide fabric can be made in an hour, making it the fastest method of fabric production.

Warp-knitted fabrics can be quite firm, like a woven fabric, or stretchy. They do not ladder or unravel and can be cut to shape when making products. They are mainly used for leisurewear, including swimwear, and underwear. Linings, lace, ribbons and other trims, net curtains and bedclothes can also be made from warp-knitted fabrics, as are industrial textiles, including geotextiles.

Examples of warp-knitted fabrics include tricot, locknit, velour and fleece.

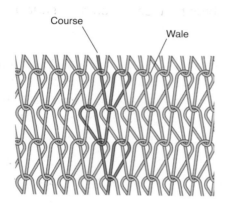

Figure 2.27 The structure of warp knitting

Weft knitting

Weft knitting is made from one long length of yarn. The loops are interlocked with those in the row above and below. All hand knitting is weft knitting, but it can also be produced on a machine.

These fabrics are stretchy, and the loops in the fabric trap air, making them warm to wear. Plain weft knits are known as 'single jersey' and tend to curl at the edges. Double jersey is made using two sets of needles and two sets of yarn, making a thicker, more stable fabric. This is easier to work with, but is not as stretchy.

The weight of the fabric depends on the type of yarn used to make it. Weft knits can be made from any kind of yarn, including fancy yarns, made from any fibres. A huge variety of textures and characteristics can be created in weft-knitted fabrics as hundreds of different stitches can be used, and the size of the loops can also be varied. Coloured yarns can be used to create patterns and designs in the fabric.

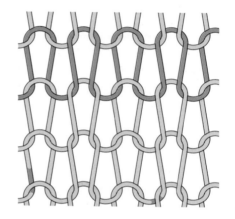

Figure 2.28 The structure of weft-knitted fabric and weft-knitted products

Weft-knitted fabrics can be made as straight lengths on flatbed machines, but as they ladder if a thread is broken or cut, products made from them need secure seams.

Alternatively, the fabrics can be produced as shaped pieces ready to be sewn together to make a product. Weft knits can also be produced as a 'tube', which is useful when making products such as socks and tights, when a seam would be uncomfortable. These fabrics are made on circular knitting machines, which are popular in factories, as they take up less room and are faster than flatbed machines. A computer can be used to control either type of knitting machine, making them easy to set up and quick to adapt to a different design.

Weft-knitted fabrics are particularly suitable for garments that need to stretch or be warm to wear. This includes T-shirts, polo shirts, sportswear, skirts, jumpers and cardigans, hats, scarves and leggings.

ACTIVITY

Make a collection of different types of knitted fabrics. Mount a small sample of each on paper. State whether it is warp or weft knitted and suggest a use for each.

Non-woven

Not all fabrics are made from yarn. Some are made directly from fibres, making them economical to produce and use.

The first stage in making a non-woven fabric is to make a 'web' of fibres. The web is held together either by tangling the fibres together or by gluing them with adhesive.

There are a number of ways of creating the web, and the way it is made affects the final performance characteristics of the fabric

produced. A 'random laid' web is created by blowing air through the fibres, causing them to fall in a random order. The fibres can be combed so that they lie straight and side by side neatly, known as 'parallel laid'. Another layer can be arranged on top of this at right angles to it, known as 'cross laid'. Manufactured fibres can be pumped directly on to a conveyor belt and are known as 'spun bonded'.

Wool felts

Wool fibres have scales on their surface, which under certain conditions will become tangled together, causing the fibres to stick together to form a fabric. The web of wool fibres is treated with an alkaline solution and heated. Pressure and a mechanical action on the web cause the fibres to tangle, creating the fabric.

Wool felts are good insulators, as air is trapped in the web and in the scales on the wool fibres. They do not fray, as they are not made of yarns, and as they have no straight grain, pattern pieces can be arranged economically on the fabric. They can also be moulded into shapes for hats and are soft. Wool felts can be made from recycled waste fibres, so are environmentally friendly. These fabrics have little elasticity and pull out of shape and tear easily. They are not strong and are not suitable for garments that require some 'give' in the fabric. They are used for insulation materials, furnishings, hats and toys.

Needle felts

These can be made from any fibre. The web is formed and then repeatedly pierced with hot barbed needles. The barbs on the needles drag fibres through the web, almost stitching it together, rather like an embellisher machine.

If the fibres are synthetic, the heat from the needle also melts the fibres, causing them to stick together.

These fabrics are relatively lightweight compared to other non-woven fabrics and slightly more elastic. As with other non-woven fabrics, they have no grain, do not fray and are cheap to manufacture. They are used for interfacings and wadding, upholstery, floor and mattress coverings, filters and dusters.

Bonded webs

The web for these fabrics is made in the same way as for other non-woven fabrics and can be bonded (or stuck) together in a number of ways. A binder or bonding agent can be put on to the web to glue it together. A chemical solvent can be added to cause the fibres in the web to soften and stick where they touch. Special fibres can be included in the web that will melt or dissolve to stick the fibres together. If synthetic fibres have been used to make the web, they can be melted with heat and then pressure applied to stick them together in small areas over the web.

Bonded webs do not fray, stretch or give. They are permeable to air and have good crease resistance, but are not as strong as woven or knitted fabrics. The lack of grain makes them economical to use, as the pattern pieces can be fitted closely together, saving fabric.

Bonded webs are used as interfacings and interlinings. They can have a fusible adhesive applied to one side, which can be melted with the heat from a domestic iron and stuck to another fabric. This product is known as 'iron-on interfacing'. Bonded webs are cheap to manufacture and are therefore often used for disposable items such as cleaning cloths, hospital items and disposable underwear. They can be impregnated with chemicals for cleaning, or antiseptic for wound dressings.

ACTIVITY

Make a collection of non-woven fabrics. Mount a small sample of each on paper and suggest a use for each.

2.4 MIXING AND BLENDING OF FIBRES AND YARNS

LEARNING OUTCOMES

In this section you will learn about:
- why fibres and yarns are blended and mixed
- how they are blended and mixed.

Fabrics are rarely made from a single type of fibre. A study of the labels found in textile products will show the percentages of the fibres combined to make the fabric used for the product. Mixing and blending fibres combine the performance characteristics of the fibres used, balancing out the good and bad points of each and producing a fabric with the performance characteristics needed by the product. It may also reduce the cost of the fabric.

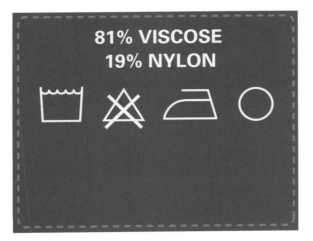

Figure 2.29 A product label showing the fibres used to make it

Mixing fibres in a fabric is done by making the warp yarns from one fibre and the weft yarns from another. The yarns are then woven to make the fabric. A blend is created by mixing the fibres together before they are made into yarn. The yarn is then woven or knitted to make the fabric.

Polyester and cotton fibres are frequently blended for items such as shirts and blouses, as well as bedding. The cotton fibres are absorbent and comfortable next to the skin, while the polyester fibres increase the abrasion resistance of the fabric, reduce the drying time and improve crease resistance.

Wool and acrylic fibres can be blended to make socks. The wool fibres are warm and comfortable next to the skin, and the acrylic fibres increase durability and make the socks easier to wash.

▌ Laminating and coating fabrics

A laminated fabric is made by joining two or more fabrics together to combine their performance characteristics. They are held together by adhesive, rather like attaching iron-on interfacing, or by heating a polymer film or foam on one side and pressing it on to the fabric it is to be joined with. GORE-TEX® is an example of a laminated fabric where a hydrophobic membrane is attached to another fabric to make it waterproof.

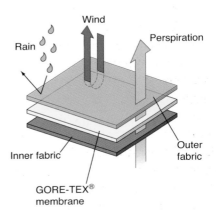

Figure 2.30 GORE-TEX®, a laminated fabric

A coated fabric has a layer of polymer applied to one side of the fabric and then it is 'fixed' in a curing oven. Fabrics are usually coated with polyurethane or polyvinylchloride (PVC). The most well-known coated fabric is probably PVC-coated cotton, used for aprons, table coverings, waterproof clothing and bags. If used for clothing, the coating should be permeable to air and moisture vapour to make it comfortable to wear. Other uses of coated fabrics include sportswear, protective and work clothing, shoes, high-visibility clothing, furniture and seat coverings, shower curtains, and many more.

ACTIVITY

Look at the labels in a range of textile products. Sketch or photograph each product and write down the fibres used to make it, including the percentages of each. Explain why those fibres have been chosen for that product.

Figure 2.31 A PVC-coated cotton shopping bag

2.5 THE FINISHING PROCESSES APPLIED TO FIBRES AND FABRICS

LEARNING OUTCOMES

In this section you will learn about:

- the various finishes applied to fabrics
- why those finishes are applied
- the advantages and disadvantages of each.

The performance characteristics a fabric has depend on a number of factors: the fibre used to make it, how the fibre is made into a yarn, and how the yarn is made into fabric. Sometimes it is possible to achieve the characteristics required in a fabric by changing some of these factors. Another way to improve the performance characteristics of a fabric is to apply a special finish to the fabric.

There are two types of special finishes: those achieved by a physical action on the fabric, known as mechanical finishes, and those achieved by adding a chemical.

EXAMINER'S TIP

In the examination you may be asked to suggest a finish for a fabric used for a particular product and give reasons for your choice.

Mechanical finishes

Brushing

Brushing raises the hairs on the surface of a fabric to make it fluffy, soft and warmer to wear. Fabrics made from cotton fibres or polyamide fibres can be given this treatment. The disadvantage is that the fabric can be weakened by the action of the metal brushes on its surface. Also, as more air is trapped in the fabric, flammability is increased. Repeated washing and ironing of the fabric will flatten the fibres and reduce the effect.

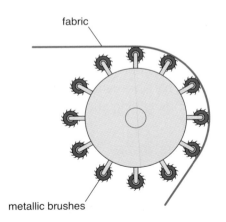

Figure 2.32 Brushing

Calendering

This is the opposite of brushing. The hairs on the surface of the fabric are flattened, making it smooth and shiny. The fabric is passed between a series of rollers that apply

pressure to the fabric to smooth it. Different effects can be achieved on the fabric depending on the surface of the rollers, the temperature of the rollers and the speed with which they turn. Moiré fabric has a 'watermark' pattern which is produced when the roller has a 'ribbed' rather than smooth surface. A pattern can be engraved on the roller, which will produce a relief pattern on the fabric, known as emboss.

The main disadvantage of calendering is that it is not permanent. Washing and ironing the fabric will affect the finish.

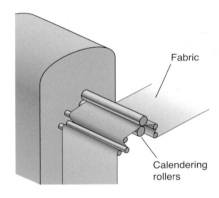

Figure 2.33 Calendering

Shrinking

When fabrics are made, they are pulled and stretched and put under tension. When they are washed for the first time, they relax and shrink. This can mean that products such as clothes do not fit after the first wash. Fabric manufacturers can pre-shrink fabrics before selling on to product manufacturers. This can be done by steaming the fabric, then laying it out on a vibrating table to relax and dry.

Pleating

Manufactured or synthetic fabrics are thermoplastic and can have pleats permanently set into them. The fabric is heated, then pleated in the desired way and held in place while the fabric cools. The pleats will remain in the fabric as long as they are washed at the correct temperature.

Cotton and viscose fabrics can be permanently pleated, but a resin needs to be applied before they are pleated and then the fabric cured in an oven to set the pleats in.

▶ Chemical finishes

These finishes involve the application of a chemical to improve the performance characteristics of the fabric.

Water repellent

One method of making fabrics water repellent is to spray the fabric with silicones. This is not always permanent. Alternatively, a fluorochemical resin is applied to either the right or wrong side of the fabric to make it waterproof and windproof. The finish can be applied to any fabric, but is particularly appropriate for all-weather wear, bags, shoes, tents and products used outside. Teflon® and Scotchgard™ are examples of water-repellent finishes.

Stain resistance

This finish can be applied to any fabric, but is particularly useful for all-weather wear, furnishing fabrics and floor coverings. A stain-resistant finish also adds some waterproof qualities. A fluorochemical resin is applied to the fabric through an environmentally friendly process that uses no chlorine or CFCs. The finish will biodegrade eventually, so disposing of the fabric is also environmentally friendly.

Teflon® and Scotchgard™ are examples of stain-resistant finishes.

Figure 2.34 A tent would benefit from a Teflon® or Scotchgard™ finish to make it water and stain repellent

Anti-static

Manufactured or synthetic fibres can build up an electrostatic charge, often known as static electricity. This causes fabrics to cling to the body and attract dust and dirt. Chemicals can be applied to the fabric to encourage the fibres to absorb moisture from the atmosphere, improving surface conductivity and reducing the build-up of static. This treatment is particularly appropriate for floor coverings and clothing such as underwear and lingerie.

Flame resistance

This finish can be applied to any fibre or fabric, but is particularly important for children's nightwear and furnishing fabrics made from cotton and viscose fibres. A chlorine or phosphorus finish is applied and fixed to the fabric to reduce flammability. The disadvantages of this finish are that it makes the fabric stiffer and reduces its strength. It is an expensive finish to apply and becomes less effective the more times the fabric is washed. Proban® is a flame-resistant finish.

Figure 2.35 Furnishing fabrics benefit from a flame-resistant finish, and carpet benefits from an anti-static finish

Hygienic

These finishes work by preventing the growth of microbes. Antimicrobial chemicals such as triclosan are applied to the surface of the fabric, or incorporated into the fibre itself to make it even longer lasting and more resistant to washing. These finishes are more effective if the fabric is breathable, as moisture aids the growth of bacteria. The finishes control odours, reduce the risk of skin irritation and infection, and prolong the life of the fabric.

Hygienic finishes can be used to protect garments, shoes, socks, sportswear and fabrics used for medical purposes.

Figure 2.36 Fabrics used for medical purposes benefit from a hygienic finish

Rot proofing

This finish protects the fibres and fabrics from organisms that destroy organic substances such as natural fibres. The finish is often applied to 'technical textiles'. Technical textiles are manufactured for their functional performance characteristics rather than aesthetics. They are used for protective clothing, furnishings, buildings, civil engineering, sport and leisure products, argricultural products, medicine and health care.

Anti-pilling

Pilling is the formation of small 'bobbles' on the surface of the fabric, which make the fabric look shabby and old. One way to prevent this is to treat the fabric with a polymer or a solvent. A new technique has been developed to treat cotton fabrics using an enzyme that removes protruding fibres before the fabric is coloured. An enzyme treatment has also been designed for wool fabrics.

Reducing the tendency for fabrics to 'pill' keeps them looking good for longer, therefore extending the life of the product.

Easy care

Cotton and viscose fabrics can be given an easy-care finish to make them dry faster and make them smoother so that little or no ironing is needed. A resin is applied to the fabric, which is then 'cured' by heat. The disadvantage of this finish is that the strength and abrasion resistance of the fabric is reduced.

Anti-felting

Wool fibres need to be washed carefully as excess heat and agitation can cause the scales on the surface of the fibres to become tangled, resulting in the fabric shrinking and becoming hard. There are two ways to reduce the likelihood of this happening. One method is to soften the tips of the scales on the fibres by giving them an oxidative treatment. The other is to coat the fibre with a synthetic polymer film. A Teflon® coating can be applied to the fibres.

ACTIVITY

For each finish described in this section, identify a product it could be applied to. Either draw the product or find a picture of it. Explain why the finish is appropriate for that product. Try to use a different product for each finish.

2.6 DYEING AND PRINTING

LEARNING OUTCOMES

In this section you will learn about:
- the various methods of colouring fabrics
- the effects created by each method.

Colour can be introduced at any stage of fabric production. Manufactured or synthetic fibres can be dyed while they are in liquid form, before passing through the spinneret. This is known as spin dyeing. Stock dyeing is when natural fibres are dyed before being spun into yarn. The yarn can be dyed before being made into fabric, or the fabric can be dyed, known as piece dyeing. It is even possible to dye a finished product. There are advantages and disadvantages to introducing colour at the various stages, and the method chosen will depend on the needs of the user, the cost, and the fibre or fabric being dyed.

EXAMINER'S TIP

In the examination you may be asked to explain how to colour a fabric using one of these methods. You will need to know the industrial method, and the method that would be used for a 'one off' or in the classroom. You will need to use notes and diagrams in your explanations.

Preparation

Before any dyeing takes place, careful preparation needs to be done. Fibres, yarns or fabrics need to be cleaned, possibly bleached, and faults corrected. Cotton or linen may be mercerized – treated with caustic soda under tension – to make the fabric dye well and improve the lustre.

Dyeing fabric

A dye bath is a liquid, usually water, which contains a pigment and a mordant. The pigment colours the fabric and the mordant is a chemical that binds the pigment permanently to the fabric. The first pigments came from plants and animals and were known as natural dyes. A limited number of colours could be produced and it was very difficult to achieve exactly the same colour a second time. The mordant used could also have an effect on the final colour.

In the nineteenth century, synthetic dyes based on chemical formulas were developed. The colours produced by these dyes were more vibrant, could be reproduced more accurately time and time again, and made a wider range of colours available.

A fabric is dyed an even colour by placing it in the dye bath for a specified amount of time, allowing the dye to come into contact with all parts of the fabric for the same amount of time. It will then be removed and dried. The colour may need to be fixed by the application of heat or a fixing agent.

Batch dyeing

This process involves dyeing a specific amount of fabric in a machine containing a set amount of dyestuff. Fixing the dye to the fabric may be done in the same machine, or may be done as a separate process. There are three main ways of ensuring an even coating of dye.

- The **jigger** system involves passing the flat fabric backwards and forwards thought the dye bath from one roller to another. It is used for medium- to heavyweight woven fabrics.

- The **winch** system pulls the fabric from the front of the dye bath to the back in a circular movement and is most suitable for lightweight woven fabrics and knitted fabrics.

- The **jet** process moves the fabric through the dye, using high-pressure jets of the dye itself.

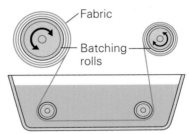

1. Jigger

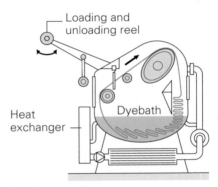

2. Winch (beck)

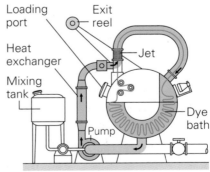

3. Jet dyeing machine

Figure 2.37 The three methods of batch dyeing

Continuous dyeing

Continuous dyeing is also known as pad dyeing. The flat fabric is passed round a roller in a relatively small dye bath and then through two rubber-covered rollers above the dye bath. These rollers squeeze the fabric and ensure even distribution of the dye before the colour is fixed into the fabric.

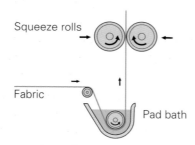

Squeeze rolls

Fabric

Pad bath

Figure 2.38 Continuous dyeing

Discharge printing

This process starts with a plain dyed fabric. A 'discharge paste' is printed on or applied to the fabric to remove the colour in certain areas to create a pattern. It is rather like applying bleach to a coloured cotton fabric to return it to its original colour.

Resist printing

Tie-dye and batik are forms of resist printing (see Chapter 3). In commercial production, a resist paste is applied or printed on to the fabric to prevent certain areas absorbing the dye. The fabric is then dyed. If only one colour is to be used, the paste will be removed. However, more paste can be added and the fabric dyed again to build up a series of colours.

Transfer printing

This method involves printing the reversed design on to special transfer printing paper first, then pressure and heat are used to transfer it from the paper on to the fabric. In industry a heated roller passes over the paper with the fabric under it. The dye becomes a vapour, which diffuses into the fabric. This is known as sublimation printing.

This type of printing can be done on a small scale using a computer and an ink-jet printer. The design is produced on the computer and reversed. The transfer printing paper is then put into the printer and the design printed on to it. The 'ink' side of the paper is placed in contact with the right side of the fabric and either a heat press or an iron is used to transfer the design. This type of printing works best on manufactured or synthetic fibres.

An even simpler way is to use transfer printing inks or transfer printing crayons. The reversed design is drawn on to paper and coloured in using one of these media. The paper is then placed ink side down on to the right side of the fabric and pressed with a hot iron to transfer the design.

Roller printing

In this method of printing, the reversed design is engraved on to a copper roller in relief. One roller is needed for each colour in the design. The maximum width of the design is the width of the roller, and the maximum repeat is the circumference of the roller. The rollers are expensive to produce, so this method of printing is only used when large amounts of fabric are to be printed.

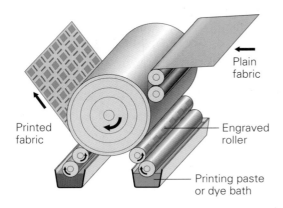

Figure 2.39 Roller printing

The roller is coated with dye and rolled over the fabric. The dye will then be fixed to make it permanent.

▶ Screen printing

This is the most popular method of printing designs on to fabric. The screen is a wooden frame with a fine mesh fabric stretched over it. The dye is moved across the screen and forced through the mesh fabric using a squeegee. The screen needs to be large enough to accommodate the design.

For one-off prints, areas of the screen can be blocked off, using paper stencils.

The stencil is a piece of paper with a hole cut in it, forming the design. One stencil is needed for each colour. The stencil is placed between the screen and the fabric being printed on. The paper prevents the dye from reaching the fabric, only allowing it through the hole. The different stencils need to match up to create the design.

If more than one print of a design is to be produced, areas of the screen need to be blocked off permanently, and this can be done in a number of ways. The screen can be coated in a light-sensitive chemical, and the areas that need to be blocked off are exposed to ultraviolet light, making the chemical insoluble. The unexposed chemical can then be washed off. Alternatively, an insoluble polymer is applied to the screen, then areas etched away to allow the dye through. Whichever method is used, one screen is needed for each colour in the design.

There are three main ways of screen printing in industry: carousel, flatbed and rotary. Carousel screen printing is often used when printing T-shirts.

Figure 2.40 Stencils and screen printing equipment

Figure 2.41 Carousel screen printing

The screens needed to print the design are attached to a revolving frame, which rotates over printing areas with the T-shirts fixed to them. The screens need to be aligned carefully when setting up to ensure the colours match when the printing is done.

Rotary screen printing is the most frequently used method, as it allows continuous production. The screen is formed into a cylinder with reservoirs to hold the printing ink. The ink is pumped through inside the screen and pushed through with a squeegee or a blade as it rolls over the fabric.

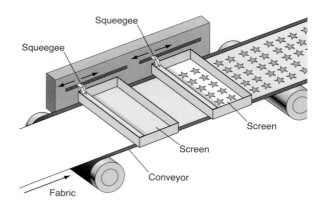

Figure 2.43 Flatbed screen printing

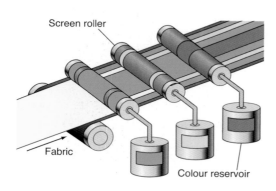

Figure 2.42 Rotary screen printing

In flatbed screen printing, the fabric is laid out flat on a conveyor and held in place. The screens are lowered on to the fabric and the printing ink applied. The fabric is moved along, a screen width at a time.

ACTIVITY

1. Try out some of the methods described in this section. Mount the samples produced and next to each one, explain how to do it.

2. Carry out research to find out more about the industrial processes described, and if possible find examples of products made using these methods.

2.7 MODERN AND SMART MATERIALS

In this section you will learn about:

- modern materials
- smart materials
- interactive materials
- nanotechnology
- new developments.

Smart and modern materials are constantly evolving, changing and developing. You will need to know what materials are being used and what ways in which they can be used and the advantages of them. Most smart and modern materials have specific functions that are used for a variety of purposes. You will need to be able to understand their structure, function and performance qualities. This area of textiles will be tested in the examination papers and can also be referred to in your design-and-make units.

Modern materials

'Modern materials' is a term used to refer to materials with special properties such as Kevlar®, Nomex®, neoprene and Polartec®. These are all trade names and have all been developed within the last 75 years. Advances in science and engineering have been taken up by fibre and fabric manufacturers and have revolutionised the fabrics now available.

Figure 2.44 Kevlar® bulletproof vest

Textiles are constantly developed with new or improved functions and this is a growing market in an important industry.

Kevlar®

Kevlar® was developed in 1965 by the company Du Pont. It has many uses and can be found in a range of products from motorcycle clothing, racing sails, body armour such as bulletproof vests and face masks to fire-protection gear. It is strong, wet or dry, and resistant to high temperatures, though it is best used with a waterproof finish as it weakens when wet. The latest developments using Kevlar® and new electronic technology are metal detector gloves. They are stab-proof and allow policemen to scan individuals for weapons. If metal is detected the gloves start to vibrate inside the wrist area. These gloves are currently being tested and worn by a number of Scottish police officers.

Figure 2.45 Nomex® fabric balloon

Nomex®

Like Kevlar®, Nomex® was developed in 1967 by Du Pont. It is used as a fabric wherever resistance from heat and flames is required. Typical uses include firefighters' uniforms, racing-car drivers' protective clothing from gloves and hoods to all-in-one underwear suits. It is also to be found in many uniforms, from airline pilots to tank drivers, and is typically paired with Kevlar® to ensure that seams are strong.

Figure 2.46 Neoprene gloves

Neoprene

Neoprene is another Du Pont invention and was created in 1930. It has a wide variety of uses and it is particularly well known for its use in sportswear, for example wetsuits, hoods, boots and gloves, and also its use in orthopaedic braces, for example wrist supports and knee supports. It is a synthetic rubber fabric.

Elastane fibres

Elastane fibres, discovered in the 1930s but not really used until Du Pont created Lycra® in the late 1950s, are another example of modern materials. Elastane fibres are best known by their trade names of Lycra® and Spandex. They are well known for their great tear resistance, durability and ease of care. Elastane fibres are used in all areas of textile production where a high performance of elasticity (stretch) is required:

- sportswear
- underwear
- swimwear
- woven and knitted fabrics.

Elastane fibres can be used with both natural and synthetic fibres, and today, without realising it, most of us wear garments that contain a percentage of elastane, from denim jeans that allow easier movement and better fit to socks that do not fall down.

ACTIVITY

Look at a selection of different garment labels and note down the percentage of elastane found in each garment.

Polartec® fabrics

Polartec® fabrics are reasonably new, having been invented in the 1970s. These fabrics are lightweight fleeces that trap body heat yet also allow body moisture to escape. These fabrics can be used with a variety of other fibres and finishes, for example odour resistant, flame resistant, durable water repellence and sun protection technology. Fleeces can come in a range of grades to distinguish the insulation and wind resistance strength.

Breathable

Breathable, waterproof and windproof fabrics such as GORE-TEX® are created by a membrane layer that is laminated to a fabric's inner surface or inserted between two layers of fabric. These fabrics are highly specialised

Figure 2.47 GORE-TEX® logo

KEY POINT

- You will need to try and keep up to date with any new materials that are developed. Newspapers and websites are useful tools to help you do this.

and are frequently used for outdoor activity clothing, including footwear. WINDSTOPPER® is another fabric that is equally durable and popular and is both windproof and breathable but not waterproof.

▶ Smart materials

What are smart materials? They are technological developments in fibre and fabric production which produce new materials, which in turn are called **smart** textiles. They can also be referred to as interactive or intelligent textiles. They are developed as a result of textile designers, scientists and technologists all working together. These smart materials can incorporate chemicals, change colour, conduct electricity, perform computational operations, collect and store energy and have many different uses and specialised applications.

Sense and react materials

Sense and react materials can be put into three groups.

- Passive smart: where only the environmental conditions are **sensed.**
- Active smart: where the material can **sense** and **react** to environmental conditions.
- Very smart: where the material **senses, reacts** and **adapts** to the environmental conditions.

These materials can be used in a wide-ranging variety of products.

Sense and react materials are designed for a specific function. For example, fabrics can be designed for swimmers that help protect from exposure to UV radiation, have anti-allergy qualities and are absorbent and antibacterial.

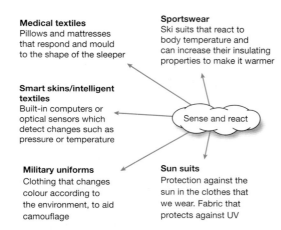

Medical textiles
Pillows and mattresses that respond and mould to the shape of the sleeper

Sportswear
Ski suits that react to body temperature and can increase their insulating properties to make it warmer

Smart skins/intelligent textiles
Built-in computers or optical sensors which detect changes such as pressure or temperature

Sense and react

Military uniforms
Clothing that changes colour according to the environment, to aid camouflage

Sun suits
Protection against the sun in the clothes that we wear. Fabric that protects against UV

Figure 2.48 Mind map of sense and react materials

Today safety garments can sense when a person is getting too hot, they can help with the absorbency of moisture and sweat and have reflective properties that enable them to be highly visible in a dangerous situation.

Medical textiles are a key growth area too, with bandages that can help heal by reacting to infections and change in temperature, indicating when they need to be changed.

Chromatic textiles are textiles that change colour according to the wearer and the environment. Dyes that are printed on to cotton or high-cotton-mix fabrics respond to changes in light and heat. A good example is novelty garments such as T-shirts with a screenprinted logo. The screenprinted logo on the T-shirt reacts to heat and thus changes colour. The dye reaction to the heat can be set at 28 °C, 30 °C or 32 °C. The finish on a T-shirt will only last for approximately five to ten washes, as it fades with washing.

Reflective inks
Minute glass balls in the ink are embedded in the fabric by being screenprinted on to the

fabric surface. The glass balls then reflect the light back to the viewer. A good example of clothing using these inks would be that worn by the emergency services.

Performance characteristics of reflective inks are:

- reflect light
- highly visible in the dark
- reflective under water
- washable.

Phosphorescent materials
Phosphorescent textiles 'glow in the dark' and are created by smart dyes that are either screenprinted or woven into fabric. Products using these include protective clothing and novelty items such as jeans, T-shirts and tops with eye-catching prints and patterns that change when under ultraviolet light.

Performance characteristics of phosphorescence are:

- colour fastness
- non-toxic
- high visibility
- non-hazardous
- washability.

In the United States a designer has developed a raincoat that glows when it rains! The 'Puddlejumper' coat is a luminescent, nylon raincoat that has sensors within it that are wired via interior electronics to electroluminescent panels. When water hits the sensors the coat lights up to create a flickering pattern that mirrors the rhythm of the rainfall.

Microencapsulation
This is the process where microscopic

capsules are applied to either a yarn or a fabric surface. Microscopic capsules can be made to contain beneficial substances, such as vitamins, oils, antiseptics, aromatic chemicals, moisturisers and antibacterial chemicals. They are released gradually through abrasion when in direct contact with the skin.

Aromatic textiles are a particular area of growth in microencapsulation. Aromatic textiles are scented products that give off an aroma when the surface of the product is rubbed. Novelty items, like T-shirts with a screenprinted image, can give off a smell such as chocolate. This effect will last for quite a time and the T-shirt can be washed as normal.

Listed below are other examples:

- **Underwear**: fabrics are scented with fruity smells, lavender calming scents and moisturising oils. Masculine scents are also used for men's underwear with smells such as musk and sandalwood.

- **Medical industry**: antiseptic capsulated products such as wound dressings, bandages and medical stitches, and medical and hospital garments to absorb moisture.

- **Children's wear**: vests for children with sensitive skins, encapsulated with moisturisers and oils. Swimwear encapsulated with sun block to prevent sunburn.

- **Sportswear**: socks and sports clothing that repels odour and fungal infections.

- **Household textiles**: bed linen encapsulated to aid relaxation and rest. Lavender and camomile to give off a soothing aroma.

Interactive materials

Electronic textiles are where electronic technology and computers are combined to produce materials that can communicate and work with us. They are produced by using conductive fibres that are developed from materials such as carbon, silver and steel. A percentage (as little as 10 per cent) of conductive thread can be used to make the product. This conductivity enables communication and interaction between the product and its user. Conductive inks have also been developed which allow a pattern to be printed on to a fabric surface which can then be used to activate the electronics.

- **Clothing**
  A jacket has been developed through a partnership between Levi's® and Philips, called Levi's ICD+. The jacket has a communications system in it which allows the wearer to listen to a phone call or to music with the same headphones by using the control panel built into the jacket. Clothing is currently being developed that could help parents keep track of their children via a camera incorporated in the item of clothing which would then use technology such as the global positioning system (GPS), which uses satellites, to pinpoint the whereabouts of the child.

- **Industrial**
  Wearable computers for workers that would be voice-activated, leaving hands free for work.

- **Medical**
  A tiny camera in a paramedic's headwear to enable visual information to be sent directly to the hospital and a doctor to respond straight away with immediate treatment.

- **Entertainment**
Club wear that could react to heat, lights and music. Skirts, shirts or trousers that have fibre optics woven into the fabric that enable the garment to 'light up'. Garments that have panels in them so that when the user dances or moves fast the panels will light up.

- **Sportswear**
Clothing that would record your activities, monitor your progress, and analyse your performance whilst playing you mood-enhancing music. Clothing for motorbiking, racing and space suits can all have electronic components within them

- **Automotive and transport**
This is the largest area for technical and electronic textiles. Used in racing cars, aircraft and space shuttles. Today we can find cars with control panels to activate heated seats, seatbelts and airbags.

A new development has seen clothing used to advertise and promote. Jackets showcased by Philips using Lumalive textiles carry dynamic advertisements and graphics that are shown by constantly changing colours. Lumalive fabrics have a range of LEDs integrated into them. Philips also plan to develop Lumalive textiles technology for fabrics used to make curtains, cushions or sofa coverings, to enable them to change colour and illuminate in order to enhance mood and atmosphere. Visit the Philips Lumalive website **www.lumalive.com** to view a demonstration of these exciting new textiles.

Figure 2.49 Philips Lumalive technology

Geo textiles

Geo Textiles are bonded or woven materials that are used in the protection and development of agricultural crops and for construction materials in civil engineering. Geo textiles have primarily been used in road construction and maintenance over the last twenty years. Geo textiles can be both synthetic and natural.

Some examples from different industries include:

- **Fishing industry**
Developments have led to Aurora luminous nets, a 'glow in the dark' net to increase night-time catch levels.

- **Building construction**
Textile 'roofs' have been built in new projects such as the Eden Project in Cornwall and the Space Centre in Leicester.

- **Canal liners**
Watertight geo membranes, which stop the drainage of water.

- **Roadways and railways**
Geo textiles act as a layer of protection, extending the life of the road or pavement.

- **Drainage and erosion**
To help control erosion and drainage, the man-made fabric acts as a filter, which allows water to pass through.

Nanotechnology

Nanotechnology is a relatively new area of the textiles market and an area that is rapidly growing and moving forwards. Nanotechnology uses revolutionary technology to enhance fabrics on a molecular level. The ability to manipulate individual atoms and place them in a desired structure has allowed new textile materials to be developed: fabric that resists spills, repels stains, wicks away moisture and can resist static without losing any comfort qualities such as feel and softness.

Nano-Tex is a company that has developed fabrics, for example Levi jeans, to become more durable, water and oil repellent, stain resistant and have a reduced need for washing, all without altering the feel of the fabric. Nano-Tex are also developing a range called Coolest Comfort which will be a range of garments that have moisture-wicking enhancements to keep consumers dry and comfortable by pulling moisture away from the body ten times faster than most normal fabrics.

Other companies have also developed areas and uses for nanotechnology such as ARC Outdoors who have developed socks with silver particles within their fibres. The silver helps provide protection against odour and fungus in the socks. Many other developments are under way and this is an area that you will need to be aware of and research to keep up to date.

- **Protective workwear/clothing**
  GORE-TEX® workwear is produced with an anti-static membrane to protect the wearer against electrostatic discharges. Water-repellent clothing technology can be used to develop improved resistance to dirt in many textile items.

- **Medical textiles**
  Bandages and dressings have silver atoms incorporated to aid the fight against bacteria.

- **Moisture-absorbing textiles**
  Kanebo Spinning Corp of Japan has produced a polyester yarn with thirty times more moisture absorption than was previously possible. The polyester has been used in the manufacture of undergarments.

Other developments using nanotechnology include fabric being produced to become more luminescent and creating a 'hue' when viewed which changes colour according to the viewpoint of the observer and the angle at which the light hits the fabric. Lighter and

Figure 2.50 ACR Outdoors nanotechnology sock

stronger fabrics are being manufactured by the spinning of nanofibres, which can then be used to produce items such as bulletproof vests.

New developments

You need to be aware of the latest developments in textiles. New ideas and techniques are being developed all the time. The **Fabrican** is one such example. Fabrican is spray-on clothing that comes out of a can. A chemical formula is sprayed directly on to the skin: thousands of fibres splatter against your skin and these fibres bind together to create a disposable garment. More development into this will lead to its uses being expanded and some potential areas of wound-healing patches and dressings are currently being explored.

Within the area of wearable electronics a lot of new products are being tested. Currently in the United States a vibrating vest is being tested that allows written messages on the back of the vest to be viewed and send warning of imminent danger. This would be beneficial when radios cannot be used and be relevant to all service people such as firefighters. Also new are fabrics that can communicate with you, that whisper messages to you. These work by either touching the fabric or wrapping it around you. Messages within the fabric can be recorded and played back.

Another interesting development is the production of Electrolux dustmate shoes – these shoes clean while you walk around your home. The shoes are made of green nylon with a flexible rubber sole and elastic unisex sock.

Other developments include '**the disappearing dress**' a environmentally friendly product which once finished with can be dissolved into a tiny amount of gel which in turn can be reconstituted into a solid once more or used to grow plants! Or the '**Hit-Air**' airbag safety system designed to protect a motorbike or horse rider's neck, spine and vital organs in the event of a fall or collision.

Many more examples can be found and you will need to research via magazine articles, books or the internet to find out more.

Figure 2.51 Fabrican spray-on dress

ACTIVITY

Some useful websites to explore are:

- www.softswitch.co.uk
- www.electrotextiles.com
- www.emtex.org
- www.nano-tex.co
- www.electrotextiles.com
- www.reflec.co.uk
- www.polaria-apparel.co.uk

QUESTIONS

1. Name and describe three smart and modern materials that are available today.

2. Describe what three factors can cause photochromic dyes to change colour.

3. Name the performance characteristics of phosphorescent dyes.

4. Explain what microencapsulation is and give an example of a current product.

5. Sportswear today uses many types of smart and modern materials. Describe as many examples as you can.

ACTIVITY

1. Look at the labels in a selection of garments and textile products and write down the fibres used to make them.

2. Make a collection of samples of fabrics. Mount them on paper and make a note of the fibres used to make them.

3. Research new developments in fibre production and make a 'fact file' about smart and modern fibres.

2.8 PRE-MANUFACTURED COMPONENTS

LEARNING OUTCOMES

In this section you will learn about:
- the use of pre-manufactured components
- the characteristics of a range of pre-manufactured components
- how to select the most appropriate pre-manufactured component.

Pre-manufactured items are items needed, other than the fabric, to make a product. These may be needed for constructing the item or to decorate it.

EXAMINER'S TIP

In the examination you will need to be able to identify the pre-manufactured components used to make a textile product.

▶ Thread

Threads can be divided into two main groups, those for use on a sewing machine and those used for hand embroidery.

Machine threads

The most commonly used thread for sewing products together is polyester cotton thread. It is strong but has some elasticity, and can be used on all types of fabric. Cotton machine thread is also available and is usually mercerized to increase its strength and sheen. It is most appropriate for cotton and linen fabrics. Silk thread is fine but strong and can be used for wool and silk fabrics. As the overall strength and durability of a product are dependent on the thread used to join the pieces together, it is important to use good-quality thread.

Buttonhole twist is a strong, thicker thread made from polyester or silk. It is used to hand-stitch buttonholes and to attach buttons. When used on the sewing machine, the settings may need to be adjusted to accommodate the thicker thread. Upholstery thread is another strong, reinforced thread designed for use in upholstering products. Metallic thread is also available and can be stitched by hand or machine, although the machine settings may need adjustment. Invisible thread is a clear, fine, strong thread made from polyamide designed for use with light- to medium-weight synthetic fabrics. Both the metallic and invisible threads are heat-sensitive, so care must be taken when pressing and ironing to avoid melting the threads.

Hand sewing threads

These are used for hand embroidery and decorative techniques. They are not designed to hold fabrics together, but can be used decoratively when working appliqué and patchwork.

Soft cotton thread is a thick embroidery thread, often used for decorative needlepoint on open-weave fabrics. It can also be used to make cords and braids. Tapestry wool has similar qualities and uses. Stranded cotton embroidery thread is the most versatile. It is made up of six separate strands, which can be separated and used individually, or in combinations to create the desired thickness. Cotton perlé thread is available in two thicknesses and has a lustrous finish.

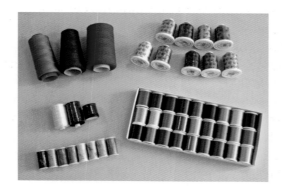

Figure 2.52 Machine threads

Figure 2.53 Hand sewing threads

ACTIVITY

Make a collection of different types of threads. Mount a small sample of each on paper and next to each sample write down the name of the thread, what it is made from and what it would be used for.

Fastenings

Fastenings are a temporary way of joining fabrics and there are many different fastenings to choose from. When selecting a fastening it is important to consider the final effect on the product, how secure the fastening needs to be, the cost, how easy the fastening will be to manufacture and how easy it is to use.

Figure 2.54 A selection of fastenings

Zips

Zips can be a decorative feature or an almost invisible fastening. They are available in a range of lengths, weights and colours, can be made from metal or plastic and can be open-ended or closed-ended. Open-ended zips separate into two pieces and are used where the two sections of the textile product need to be completely separated, for example in the front of a coat or jacket. Closed-ended zips are used to increase the size of an opening, for example at the waistband of a skirt or trousers. It is important to select the correct weight of zip for the fabric being used to make the product.

Velcro®

Velcro® is the trade name for hook-and-loop tape. The fastening consists of two layers of tape. One side of one of the tapes is covered with fine nylon hooks; the other tape has one side covered with a fluffy, looped surface. When the hooks are pressed against the looped surface, the two layers stick together, as the hooks become tangled in the loops.

This is a very easy fastening to use and is often found on items designed for children or the elderly. It can also be applied where the fastening needs to be adjustable, for example on cuffs or on waistbands. The fastening is quite secure, washable and easy to apply. Two disadvantages of this fastening are that the hooks tend to collect bits of threads and fluff, making it look untidy, and the hooks can catch on other fabrics and 'snag' them.

Buttons

Buttons come in a huge range of shapes, sizes and designs and can be used as a decorative feature. They are usually made from plastic, glass or metal. Some are designed to be covered in fabric. Flat buttons have holes punched through them allowing them to be stitched on, while others have a 'shank' formed at the back to stitch through. A button on its own is not a fastening; it needs a buttonhole or a loop to fasten through.

Hook and eye

These are traditionally made from metal and can be black or silver and are available in a range of sizes. The 'eye' can be a curved loop or a straight bar, or can be made by stitching. This type of fastening is strong and will not come undone under stress. It is a discreet fastening and is often used on waistbands or at the top of a zip.

Press studs

These come in a variety of shapes and sizes and can be made from metal or plastic. They are sometimes called snap fasteners. They are not suitable to use where the fastening will come under stress, as they pull apart quite easily. Plastic press studs are available in a range of colours as well as clear versions and are designed to be discreet. They are stitched on by hand.

Poppers

These are rather like press studs, but are attached using a special tool and a hammer. They are visible from both sides of the fabric and can be used as a decorative feature. They are made from metal or plastic and are available in a range of sizes and colours. They are not suitable for very thick fabrics or quilted fabrics, as the two parts do not always lock together properly and may fall off.

Ties

These are made from ribbon, braid or cord. They can be cut to the desired length and stitched in place by hand or machine. The cut end that is loose may need to be finished to prevent fraying.

Laces

These are available in set lengths with the ends finished to prevent fraying and make the laces easier to thread through holes. They are available in a range of thicknesses, colours and designs.

Parachute clips

These are made from plastic and most commonly available in black, although they do come in a range of sizes. They consist of two parts which push together, one inside the other, and clip in place. Squeezing the inner section and pulling the two halves apart opens them. They are very secure and are often used on bags. They are designed to be seen and are usually attached using braid, making them adjustable.

Figure 2.55 Parachute clips

Toggles

There are two different types of toggles. One type is a traditional wooden fastening, often made from plastic nowadays, which is rather like a button and is fastened through a loop or buttonhole. The other sort of toggle is used at the end of a drawstring and it usually has a locking mechanism which allows the drawstring to be pulled up and secured in place, using the toggle, and released as necessary.

Eyelets

Eyelets are made of metal and are designed to neaten the edges of small holes made in fabric. They are small 'tubes' of metal that fit inside the hole. They can be used in a belt to fasten with a buckle, or attached to products to allow laces or ribbon to be threaded through.

ACTIVITY

Collect pictures of textile products that use each of the different fastenings. Next to each picture, identify the fastening used and explain why it is the most appropriate fastening for that product. Suggest alternatives where possible and give reasons for your choice.

▶ Construction Items

These components perform a function to help the textile product perform as intended. They may improve the look of the product, but they serve a purpose.

Shoulder pads

The size, shape and thickness of shoulder pads change with fashion. They can be used to disguise figure faults, for example to build up narrow or rounded shoulders, or change the shape of the product so that it is fashionable. There are different types of shoulder pad available for different uses. Dress and blouse shoulder pads are usually covered in fabric, sometimes to match the garment. As suit and coat shoulder pads are usually placed between the garment and a lining fabric, they are not covered.

Cuffing

Cuffing is a tube of knitted fabric designed to be attached to the bottom of a sleeve. It can often be seen on sweatshirts. It is a way of finishing off the end of the sleeve which allows stretch for the hand to go through but will return to its original size, gripping the wrist and keeping the end of the sleeve in place.

Interfacing

Interfacing is used in garments to stiffen and strengthen areas. For example, interfacing is used in a collar to help it maintain its shape. Interfacing is used at the front of a blouse or shirt to strengthen and stabilise the fabric where the buttons and buttonholes are. Interfacing can also be used effectively when working decorative techniques such as patchwork and appliqué. Interfacing can be a woven or bonded fabric and comes in a variety of weights and colours.

Pockets

When a pocket is not visible from the outside of a product, it does not have to be made from the same fabric as the product. This means that it can be 'bought in'. For example, the patch pockets on the back of a pair of jeans are visible and are usually made from the same fabric as the jeans. However, the pockets in the front lie inside the jeans when they are being worn and are often made from a different fabric, often a thinner white fabric. This reduces the bulk and saves time when the jeans are made, and therefore reduces costs.

ACTIVITY

Collect pictures of textile products that use each of the different construction items. Next to each picture, identify the construction item used and explain why it has been used.

Decorative items

These are items that do not need to be used on a product in order for it to function properly. They are purely decorative and are used to make the product more visually appealing.

Figure 2.56 A selection of pre-manufactured decorative items

Appliqué motifs

These are commercially produced decorative 'patches' designed to be stitched on to products. They often have some decorative machine stitching added to create the design, and can be popular cartoon characters or logos for companies, organisations or clubs. The company manufacturing the appliqué motif will have negotiated copyright with the company who owns it. As well as being a quick method of decorating a product, they can be used to repair items, as the motif can

be stitched over a hole or thinning area of the product.

Ribbon

Ribbon is available in countless widths, colours and textures and has lots of uses in textile products. It can be made from almost any fibre, including metallic fibres, and comes in a variety of weights and widths. It can be stitched on to fabric by hand or machine.

Lace

The term lace is applied to a number of textile fabrics and components. Lace can be a fabric, made in a variety of widths and bought by the metre. Lace can also be an edging strip, available in a variety of widths, colours and designs. Lace motifs are also available, which are rather like appliqué motifs and are stitched on to fabric and lace collars can also be bought and attached to garments.

Braid

Braid is traditionally a woven decorative trim that is finished on both edges. It can be made from almost any fibre, including metallic fibres, and comes in a variety of weights and widths. Some braids are plain while others

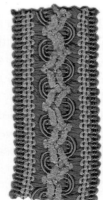

Figure 2.57 Braids

have a pattern woven into them. When you are choosing a braid, it should not be heavier than the fabric it is being stitched to.

Ric-rac

This is a wavy braid traditionally made from cotton fibres and available in a variety of colours and widths.

ACTIVITY

Collect pictures of textile products that use each of the different decorative items. Next to each picture, identify the decorative item used. Suggest alternatives where possible and give reasons for your choice.

CHAPTER 3

TOOLS, EQUIPMENT AND PROCESSES

By the end of this chapter you should have developed a knowledge and understanding of:

- The basic equipment used in textiles and its safe use, alternative tools and equipment and textiles machinery and its safety
- How to use paper patterns and the symbols used on them
- Different methods of shaping textiles
- Ways of joining fabrics
- Methods of neatening the edges of fabrics and how to choose your method
- Attaching fastenings, using decorative components and applying construction items to garments
- Ways of colouring fabrics using dyeing and printing
- Methods of adding decoration using fabrics and thread

3.1 TOOLS AND EQUIPMENT

In this section you will learn about:

- basic textiles equipment
- awareness of alternative tools and equipment
- tool selection – effective and safe use
- safety checks on machines.

You will need to know how to identify and use a range of tools and equipment, and most importantly know how to select the appropriate tool and use it safely and effectively. You will need to understand safe working practice when using tools and equipment for both personal safety and for the safety of others.

Sewing machines: From basic models to computer-controlled programmable machines

Embroidery ring: circular frame to hold fabric taut and flat

Notcher: to mark the fabric for pattern matching

Tape measure: flexible to measure around people/objects

Overlocker: sews, neatens and trims in one action.

BASIC TOOLS AND EQUIPEMENT

Pins: steel, nickel-plated or brass

Metre rule: to measure large pieces of fabric

Tracing wheel / Tailor's chalk: for marking the fabric temporarily

Unpicker: for unpicking stitches and cutting between sewn buttonholes

Needles: hand/machine in different sizes and with different-shaped eyes

Figure 3.2 A mind map of basic tools and equipment used in textiles

Figure 3.1 Textiles equipment

Basic textile equipment

When designing and making textile products, a wide variety of tools and equipment needs to be used. Understanding how they should be used helps to ensure a high-quality product and safe working practice. You need to select the most appropriate equipment for the fabric you are using.

All equipment has a specific function and this enables you to select the right equipment to use:

Hand sewing needles

These come in a range of thicknesses and lengths, and have different-shaped eyes and points. The type of needle used will depend on the task to be done, the type of fabric and the thread you have selected. All needles need to be sharp and smooth.

- **Sharps** are needles for general hand sewing such as tacking and gathering.
- **Betweens** are short needles for quilting.
- **Crewel** needles are used for embroidery and have a sharp point and large eye.
- **Ball-point** needles are used for stretch fabrics, and penetrate the fabric more easily.
- **Bodkins** are short needles with a rounded point and are used to thread cord and elastic through casings.

Machine needles

Machine needles are used for sewing machines and come in a variety of sizes. There are four main types of machine needle:

- **Ball-point** needles have a rounded end and are used on knitted and elastomeric fibre fabrics.
- **Cutting-point** needles have a tip which cuts, and are used for fabrics such as leather.
- **Set-point** needles have a sharp point which pierces the fabric and they are used on most woven fabrics.
- **Twin needles** are two needles mounted side by side, and the two threaded needles work together with the bobbin thread to create a tuck in the fabric between the rows of stitching.

The thickness of a needle is measured in hundredths of a millimetre. Size 70 or below is considered to be a fine needle and suitable for lightweight finer fabrics such as chiffon, sizes 80 and 90 are medium, and over 110 is thick.

Other sewing equipment

Pins

Available in a variety of sizes, pins can be steel, nickel-plated or brass and are used to hold fabric in place for cutting or sewing. Some have bright coloured heads to make them easier to see.

Unpickers

These are used to unpick both hand and machine stitches and to cut between the two rows of stitching which form a buttonhole. They have a very sharp point and a cutting edge.

Tape measure

Used to measure accurately and should be made from non-stretch fabric and be clearly marked in cm or both cm and mm. It should have metal ends so that it cannot fray. Tape measures are used for taking body measurements and measuring round objects.

Tailor's chalk

This is usually triangular in shape and is used to transfer temporary markings on to fabric. It can be bought in a pencil shape with a brush at the end to help erase the markings.

Tracing wheel/carbon paper

These are also used to transfer markings on to fabric. The tracing wheel leaves a fine line, with the carbon paper coming in a variety of colours and washing out easily.

Embroidery rings

These are circular frames made from wood or plastic and can be used for both hand and machine embroidery to hold the fabric taut and flat. An embroidery ring is essential when working free machine embroidery.

Metric rule

This is a large wooden or metal ruler that is used to measure out large pieces of fabric.

KEY POINTS

- Buy good-quality equipment.
- Use the equipment correctly and safely.

Cutting tools

There are many types of cutting tools that can be used in either your school or in industry. The diagram in Figure 3.3 shows a range of cutting tools that you will need to be aware of.

Paper scissors: for cutting paper and card

Dressmaker's shears: for cutting fabric

Band knife: precision cutting in industry

Circular cutter/ power shears: for cutting single-ply fabric to length. Hand process only

Craft knife: for cutting paper and card.

Pinking shears: for finishing edges

CUTTING TOOLS

Snippers: a tool with a spring loaded blade

Embroidery scissors: short blades with long points

Automated computer: controlled cutter: fully automated, accurate and fast

Unpicker: for unpicking stitches and cutting between two rows of stitching which form a buttonhole

Die cutter: the cutting dies are made in the shape of the pattern pieces and are cut out to the exact shape. Good for mass production

Figure 3.3 Cutting tools

Awareness of alternative tools and equipment.

There are many different types of tools and equipment that can be used in industry. A lot of tools and equipment can be computer-operated and computer-controlled. When looking at cutting tools, for example, it is important to refer to Chapter 5, Industrial Production, and read about the specific tools and equipment that are used in an industrial context within computerised cutting.

Joining tools and equipment

You will need to be able to identify and list tools and equipment that can be used for joining both permanently and temporarily. Each piece of equipment has advantages and disadvantages, most items of large equipment are electrically operated and many are computer-controlled.

Sewing machines

There are basic machines, automatic machines and computer-controlled machines. Each model varies, some allowing the user to design their own stitches, others supplying computer disks to extend the stitching, all resulting in permanent joining. Each machine will come with its own manual which should be studied carefully. Electronic or computer-controlled sewing machines have a vast range of programmable decorative and functional stitches. Information can be programmed into the machine and held in the machine's memory. The machine will set the stitch length and its width. Models of machines will vary and some allow the user to design their own stitches and combinations to extend the range of stitching.

Figure 3.5 Overlocker

Knitting machines

Knitting machines are used for knitting either patterned or textured fabrics. They can be computer-controlled with programmable software, and result in permanent joining.

Poem machine

An embroidery machine that is computer-controlled.

Weaving loom

There are different types of looms that are used to weave fabrics. They can be computer-controlled.

Pins and needles

A temporary method of joining fabric.

Figure 3.4 A computerised sewing machine versus a basic sewing machine

Overlockers

An overlocker has no bobbin and uses three or four cones of thread from the top. The machine stitches the seam, neatens the edge and trims away the excess fabric in one operation, thus speeding up the process. It results in permanent joining.

▌ Ironing and pressing equipment

An iron is a very important piece of equipment. It can be used for a number of tasks:

- removing creases from fabric
- pressing edges, pleats and seams
- setting dyes into fabrics
- fusing bonded fabrics together
- pressing folds in fabric
- removing wax from batik
- finishing garments.

All irons work through using heat and the pressure applied by the user. An iron's temperature ranges between 60°C and 220°C. It is vital that the correct temperature is selected for the fabric being ironed: too hot will damage and burn the fabric, too cool will have no effect.

Steam irons have a reservoir of water which is heated to make steam. The use of steam makes creases in fabric easier to remove.

Ironing boards and sleeve boards are used to provide a stable surface on which to use the iron. The boards are covered with padding and a layer of fabric, usually cotton as this is not damaged easily by heat.

▶ Equipment and tool selection – effective and safe use

Selecting the right equipment and tools for use is a key area of importance. Many potential hazards exist in a textiles workplace, whether it is a classroom or a factory. Electrical machinery, sharp handheld objects or computerised cutting tools can all cause injury. You have a responsibility to ensure that you work in a safe and controlled way and follow safety instructions. All workplaces should display details of health and safety procedures.

KEY POINTS

- You should be able to use a range of tools and equipment effectively and safely.
- Remember that in the examination marks are only awarded for the full and correct names of tools and equipment.

▶ Safety checks on electrical equipment

It is required by law that all electrical equipment, whether used in a school or industry, is maintained. The frequency of user checks, inspections and testing needed will depend on the equipment, the environment in which it is used and the results of previous checks.

Many faults with electrical work equipment can be found during a simple visual inspection:

- Switch off and unplug the equipment before you start any checks.
- Check that the plug does not look damaged and that the cable is properly secured and no internal wires are visible.
- Check that the electrical cable does not look damaged. Damaged cable should be replaced with a new cable by a competent person.
- Check that the outer cover of the equipment is not damaged in a way that will give rise to electrical or mechanical hazards.
- Check for burn marks or staining that suggest the equipment is overheating.
- Position any trailing wires so that they are not a trip hazard and are less likely to get damaged.
- Ensure you are wearing any appropriate personal protective equipment.

If you are concerned about the safety of the equipment, you should stop it being used and ask a competent person to undertake a more thorough check.

▶ Safety checks on machines

There is also a legal requirement for safety checks on all textile machinery, covered by BS EN ISO 11111:1995, British Standards Institution 1995. You do not need to know all details of this requirement, just some key areas:

- Work equipment includes all equipment including machinery and tools.

- Work equipment should be maintained in an efficient working order and state of repair. Repairs, services and modifications to machinery and tools should be undertaken by those who have been trained to do so.

- All guards and protective devices should be adequate, robust and properly maintained.

- Adequate health and safety information should be displayed in a prominent position in the workplace.

ACTIVITY

1. Try and make a list of all the tools and equipment that you have used in your textiles coursework project. Explain why each piece was chosen.

2. Compare your list of equipment with equipment that would have been used if your product was going to be commercially manufactured.

3. Produce a list of electrical safety points you would need to check before using equipment for the first time.

3.2 PROCESSES

▶ Use of patterns

LEARNING OUTCOMES

In this section you will learn about:
- the use of commercial patterns
- pattern symbols – what they mean and how to use them
- methods of transferring pattern markings.

When making textile products, long lengths of fabric have to be cut into pieces the correct size and shape to make the product. These are then stitched together to make the item. To ensure the pieces are cut accurately, paper patterns can be used.

EXAMINER'S TIP

In the examination you will need to be able to describe how to produce a pattern for a simple textile product. You will be required to show all pattern symbols and other information needed to make the product using the pattern.

Commercial patterns

There are a number of companies that produce commercial patterns for making all kinds of textile items such as clothing, toys, accessories and soft furnishings. The pattern pieces have been drawn out and tested to ensure the pieces fit together to make the product. Along with the pattern pieces, a set of instructions is provided giving details of how to make the product. Using a commercial pattern avoids the need to develop patterns and test them, saving a lot of time.

On the front of the envelope that holds the pattern pieces is a picture of the products that can be made using that pattern. If it is for a garment, there will usually be several versions to choose from. On the back is a list of suitable fabrics and the components needed to make the item. If the pattern is for a garment, there will be a size chart showing body measurements for the various sizes, and also finished measurements.

Using a commercial pattern

The pattern pieces are printed on large sheets of tissue paper. They need to be cut out and separated ready for use. The instruction sheet will indicate which pattern pieces are needed to make the product. Each pattern piece has important information printed on it to ensure that the pieces are cut out correctly and also to help assemble the product. There will be an indication of which piece it is, e.g. front, back, how many need to be cut, positions of fastenings, the size and allowances for hems and turnings. A lot of other information is given in the form of symbols.

Pattern symbols

Table 3.1 shows the most frequently used symbols and what each one means. It is particularly important to follow the instructions for cutting out. If the fabric is not cut correctly, the overall quality of the finished product will be affected.

ACTIVITY

1. **Find a commercial pattern. Make a list of the information on the front of the pattern envelope and the information on the back.**

2. **Look at the instruction sheet in the pattern envelope and compare the pattern symbols shown on it to those in Table 3.1. Make a note of any differences.**

Construction of patterns for simple products

Although a commercial pattern is useful when making a garment, patterns for simple products can be made quite easily. A 'net' of the product needs to be drawn out at the actual size the item will be when finished. Then a seam allowance is added where two pieces need to be stitched together, and allowances for hems and turnings where needed. The pattern symbols used on commercial patterns should be included, as this will help with cutting out and construction. An example is shown in Figure 3.6.

Figure 3.6 Making a simple pattern

Pattern Symbol	Meaning	Importance
	Lengthen and shorten lines	These must be adjusted before using the pattern to maintain the proportion of a garment. Fold into a pleat to shorten the length, cut and spread to increase length.
	Straight grain arrow	Must be placed parallel to the selvedge of the fabric so that the grain runs correctly and the product hangs properly or lies flat
	Place on fold arrow	The edge indicated has to be against a fold of the fabric, as the piece is symmetrical and needs to open out.
	Cutting line	This is the line to cut along. Cutting too far in will make the product smaller than intended. Cutting too far outside the line will make the item too big. In either case other pieces may not fit together if not cut accurately.
	Stitching line	This shows where the stitching should be when joining sections of fabric together. Too far in from the edge of the fabric will make the product smaller, too close to the edge will make the product bigger. The usual amount allowed for a seam is 1.5cm.
	Seam allowance	This is the distance between the cutting line and the stitching line, usually 1.5cm on a commercial pattern.
	Dot	Indicates a position. For example a dart, gathers, pleats, tucks, pocket, end of a zip.
	Notch	Indicates which pieces fit together and how they need to be aligned. Can also be used to indicate the position of gathers.
	Centre line	Indicates the centre front or centre back of a garment
	Button and buttonhole position	Shows where to work the buttonhole and stitch the button for correct spacing. These can be adjusted if required.

Table 3.1 Pattern symbols and their meanings

ACTIVITY

Choose a simple textile product and make a paper pattern that could have been used to make it. Include all pattern symbols and other information needed to make the product using the pattern.

Pattern markings – methods

When the fabric has been cut to shape, some of the information on the pattern pieces needs to be transferred to the fabric: for example, if a garment is being made, the positions of darts, pockets, buttons and buttonholes. This can be done in a number of ways.

Tailor tacks

These are stitches worked in double thread through the paper pattern and the layers of fabric underneath. They have the advantage of marking all layers of fabric in one go and can be seen from both sides of the fabric. They are loose and can pull out easily, but do not permanently mark the fabric.

Tailor's chalk or pencil

Tailor's chalk is traditionally a triangular-shaped piece of chalk that can be used to draw directly on to the fabric. It leaves a fine layer of dust on the surface of the fabric that will rub off quite easily. It can also be bought in the form of a pencil, which is easier to use. In either case, it marks only one side of the fabric. It is good for marking lines on fabric. 'Vanishing pens' are also available. These are like felt-tip pens and can be used to draw on fabrics. They 'disappear' when wetted or after a certain amount of time. These should be tested before use, as sometimes the marks reappear after a time.

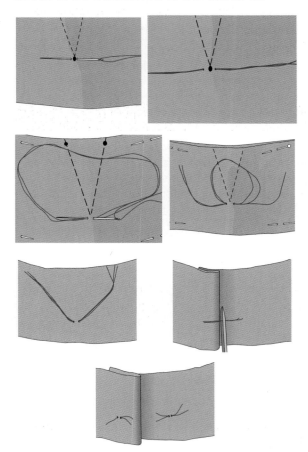

Figure 3.7 How to work a tailor tack

Figure 3.8 Tailor's chalk and pencils

Tracing wheel and carbon paper

Carbon paper is paper coated with a coloured substance that comes off when pressure is applied. The coloured side is placed next to the wrong side of the fabric. Pressure is applied using either a pencil or a tracing wheel to cause the transfer. Only one side of the fabric is marked and care should be taken that the fabric is not permanently marked.

Figure 3.9 Tracing wheel and carbon paper

ACTIVITY

Produce samples of the different methods of marking fabrics on a range of different fabrics.

EXAMINER'S TIP

In the examination you may be asked to select a method of transferring pattern marking and explain how to use it.

3.3 DISPOSAL OF FULLNESS

LEARNING OUTCOMES

In this section you will learn about:
- methods of disposal of fullness.

Many textile products are three-dimensional, for example garments and toys. The flat fabric can be shaped in a number of ways depending on what the product is, the effect required on the finished product and the fabric being used.

EXAMINER'S TIP

In the examination you will need to be able to describe, using notes and diagrams, how to work the different methods of disposal of fullness.

Darts

The position of the dart is marked by dots on a pattern piece. A triangular section of fabric is removed by a row of stitching, the wide end at the edge of the fabric tapering to a point. Darts are often used at waistlines on trousers and skirts, and at the bust of blouses and dresses. A double-pointed dart may be used in a one-piece dress, releasing fullness at the bust and hip, and taking it in at the waist. Darts give a smooth effect.

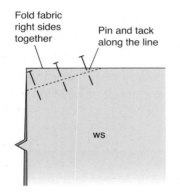

Figure 3.10 A dart

Gathers

Gathers are tiny tucks created in fabric. If the gathers are worked on a sewing machine, a longer stitch length should be used, and the top tension loosened slightly. One row of stitching is worked on the stitching line, and another one worked in the seam allowance, on the right side of the fabric. The ends of the stitching are not fastened off but left loose, so that the bobbin thread can be pulled up to gather the fabric. The stitching should be pulled from both ends of the rows of stitching until the fabric is gathered to the required amount. A pin is then put in the fabric at each end of the stitching and the

thread wrapped around it in a figure of eight. The gathers should be evened out between the pins and can be joined to another piece of fabric.

The gathering can also be worked by hand – double thread adds strength so that the thread does not snap when it is pulled up. Large amounts of gathering should be done in sections, or a ruffler attachment can be used on the sewing machine.

Gathers are often found at the waistband of skirts, sleeve heads, cuffs, on frills and yokes as well as bedding valances. They give a soft, floating effect.

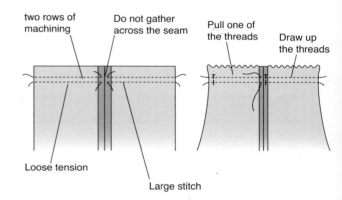

Figure 3.11 Gathering

Pleats

These are folds in the fabric, and can be either pressed in place using an iron, or stitched part way down their length. If they are made in a thermoplastic fabric (see Chapter 2) the pleats can be heat-set in place so when washed at a low temperature they remain in place. There are three main types of pleat: knife pleats, box pleats and inverted pleats.

Pleats can be found in skirts at the waistband or towards the hem to allow easier movement, at a cuff, or on a bedding valance.

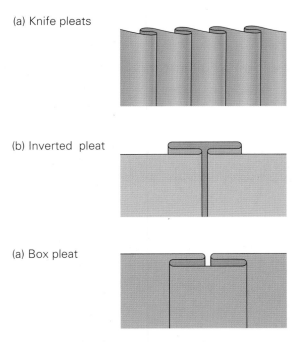

(a) Knife pleats

(b) Inverted pleat

(a) Box pleat

Figure 3.12 Knife pleats, an inverted pleat and a box pleat

Tucks

Tucks are usually narrower than pleats, are not pressed along their length. They are usually held in place by a row of stitching across the top of the tuck or by being joined to another piece of fabric. They are often found at the waistband of trousers or in the back of a shirt. Pin tucks are very narrow tucks stitched in place down their entire length and are more of a decorative feature than a method of disposal of fullness.

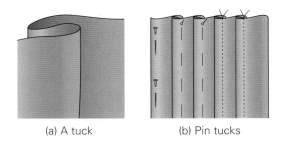

(a) A tuck   (b) Pin tucks

Figure 3.13 A tuck and pin tucks

The use of elastic in a casing

A casing is a tube formed in a piece of fabric, usually at an edge. Elastic or a drawstring can be threaded through the tube to pull the fabric in. Elastic in a casing is often used at a cuff or a waistband, or at the bottom of jogging bottoms. The fabric is pulled in, but the elastic stretches to allow more flexibility. A drawstring in a casing allows the size to be adjusted, pulled up tighter or left looser depending on requirements. This is used at the waistband of coats and jacket as well as at the top of a rucksack. Casings can also be made using bias binding (see section 3.5, on neatening edges).

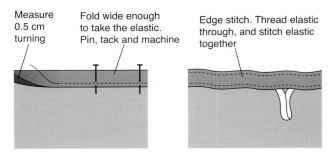

Measure 0.5 cm turning

Fold wide enough to take the elastic. Pin, tack and machine

Edge stitch. Thread elastic through, and stitch elastic together

Figure 3.14 A casing with elastic

ACTIVITY

1. Work a sample of each of the different methods of disposal of fullness.

2. Mount each sample on a piece of paper and next to it write down how to do it. Include some diagrams.

3. Find a picture of a textile product that uses each method and mount it on the paper with the sample.

3.4 JOINING FABRICS

LEARNING OUTCOMES

In this section you will learn about:
- temporary methods of joining fabrics
- permanent methods of joining fabrics.

Sometimes fabrics or components need to be held together temporarily. This may be to hold layers of fabric or components in place while a permanent join is made. Sometimes a temporary join is made to test the fit of the pieces, or the fit of the product, for example a garment, so that adjustments can be made before the join becomes permanent.

EXAMINER'S TIP

In the examination you will need to be able to describe, using notes and diagrams, how to work the different methods of joining fabrics.

▶ Temporary methods of joining fabrics

Pinning

Pins are a quick, easy way of temporarily joining fabrics and components. The pins need to be clean and sharp so that they do not damage the fabrics. They can also be used as part of the packaging of a product, to hold it in place within the packaging.

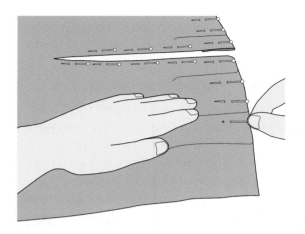

Figure 3.15 An example of pins in use

Tacking

Tacking is a temporary stitch that can be worked by hand or machine. The hand stitch is a running stitch. If a machine stitch is used, it is a longer stitch than would be used to permanently join fabrics. A cheap, slightly thicker thread is used, often in a contrasting colour so that it is easily seen, making it easy to pull out. Tacking is a more secure and less

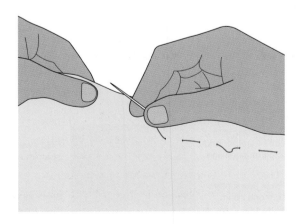

Figure 3.16 Example of tacking

bulky method of temporarily joining fabrics than pinning and is easier to machine over. A row of tacking stitches can also be used as a method of marking fabric.

Fastenings

Fastenings are also a method of temporarily joining fabrics when a product is in use. Refer to Chapter 2 for more information on fastenings.

Seams

A seam is used to permanently join two pieces of fabric together. There are a number of different seams to choose from. The factors to consider when choosing a seam are: the type of fabric being used, the position and purpose of the seam, ease of manufacture and the end use of the product.

The plain or open seam

This is the simplest seam. The seam is flat and can be used on most fabrics and products. The fabrics to be joined are placed with their right sides together, matching the raw edges of the fabric and any dots and notches. The layers are pinned together. If the pins are placed at right angles to the stitching line, they can be machined over to avoid the need to tack the layers together.

The stitching line is usually 1.5 cm from the edge of the fabric and can be marked using tailor's chalk or tailor's pencil. Alternatively, there may be a guide on the sewing machine. The seam is stitched along the stitching line using a straight stitch, remembering to reverse at the start and finish of the stitching. The pins or tacking can then be removed. The seam should be pressed open using an iron.

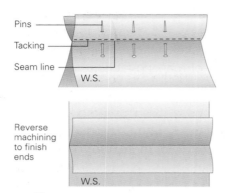

Figure 3.17 The open seam

Methods of neatening edges on the plain or open seam

The raw edges of woven fabric will fray unless neatened. The fraying not only looks untidy, but if the fabric frays back as far as the stitching of the seam, the seam will pull apart. Knitted fabrics can unravel, curl or loose their shape, so they also benefit from being neatened. There are a number of ways of neatening the seam depending on the type of fabric and the position of the seam on the garment.

The raw edges can be turned under 5 mm and stitched using either a straight stitch or a zigzag. The edge can be zigzagged without turning under, which is a less bulky finish. Bias binding can be used on the edge, or the seam can be overlocked, which is a method used in industry.

The double-stitched seam

This is a strong, flat seam often used on jeans, overalls and pyjamas. All the edges of the fabric are enclosed so they do not need neatening and the two rows of machining form a decorative feature, particularly if a contrasting colour thread is used. Depending on whether the right or wrong sides are placed together for the first join, one or two rows of stitching will be visible on the right side of the product.

If two rows of stitching are to show on the right side of the product, place the fabrics to be joined wrong sides together. Match the edges and notches. Pin and either tack on the stitching line or mark it. Use a straight stitch to machine along the stitching line, reversing at the start and finish. Remove the pins and tacking. Press the seam open. Trim one side of the seam allowance to 5 mm. Fold the other side of the seam allowance over it, tucking the raw edge under. Pin and tack in place. Machine stitch using a straight stitch close to the folded edge.

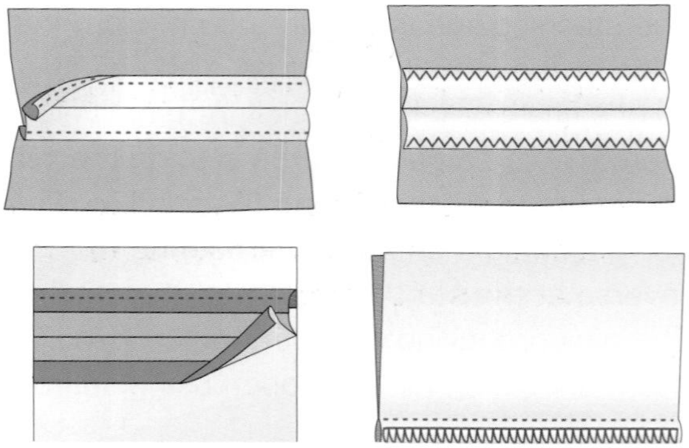

Figure 3.18 Methods of neatening the open seam

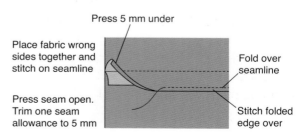

Figure 3.19 The double-stitched seam

The French seam

This seam is suitable for fine or sheer fabrics. All of the edges are enclosed so that the seam does not need neatening.

Place the fabrics to be joined wrong sides together, matching the edges and notches. Pin and tack. The first row of straight stitching is done 1 cm from the raw edges, remembering to reverse at the start and end of the row. Press the seam open and trim the seam allowance to 3 mm. Fold the fabric so that the right sides are together, bringing the seam to the folded edge. Pin and tack. Machine, again using a straight stitch 5 mm from the folded edge, reversing at the start and finish. Press the seam.

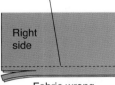
Stitch plain seam 1 cm from edge. Trim seam allowance to 3 mm

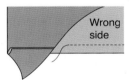
Turn to right side. Press flat Stitch exactly on the seam line 5 mm away

Right side

Wrong side

Fabric wrong sides together

Figure 3.20 The French seam

The overlocked seam

This seam can be used on almost any type of fabric and in any position on a product. The seam is made using an overlocking machine. These machines do not have a lower bobbin, but use cones of thread holding thousands of metres each on top of the machine. This reduces the need for constant rewinding of the bobbin and allows more stitching to be done before the threads run out. Also, the amount of thread remaining can be easily seen at all times. It is the most common seam used in industry (see Chapter 5).

There are various types of overlockers. Some stitch the seam, neaten it and trim off the excess fabric, while others neaten and trim off the excess fabric. In either case, the time taken to stitch and neaten the seam is reduced. The overlocking also has more 'give' than a seam stitched using a straight stitch, allowing a certain amount of stretch, making it particularly useful for knitted fabrics as it reduces the risk of the seam threads snapping when the product is in use.

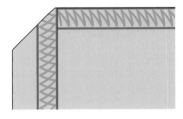

Figure 3.21 An overlocked seam

ACTIVITY

1. Work a sample of each of the different seams.

2. Mount each sample on a piece of paper and next to it write down how to do it. Include some diagrams.

3. Find a picture of a textile product that uses each seam and mount it on the paper with the sample.

3.5 NEATENING EDGES

In this section you will learn about:
- the methods of neatening the edges of fabrics
- the effect each method gives
- how to choose the most appropriate method.

Sometimes the edges of the fabric used to make a product are not incorporated into a seam. These edges need to be neatened to improve the look of the product and prolong its life. There are a number of methods that can be used depending on the type of fabric, the position of the edge on the product and the final effect required.

EXAMINER'S TIP

In the examination you will need to be able to describe, using notes and diagrams, how to neaten the edges of fabrics.

Bias binding

Bias binding is a strip of fabric cut on the bias of a woven fabric (see Chapter 2) giving the strip a slightly stretchy quality. The edges are folded to the wrong side and pressed in place. Bias binding can either be made to suit the product or be bought as a manufactured component. The stretch of the binding means it can be applied to curved edges and a smooth finish achieved. It is usually applied so that it is visible on both sides of the edge being neatened, giving a decorative edging; however, it can be attached so that it can only be seen on one side of the edge. Bias binding can be used to neaten the edges of seams, armhole and neck edges on garments, the edges of quilted products, play mats, and can also be used to create a casing.

Figure 3.22 Uses of bias binding

There are several ways to apply bias binding to an edge. The quickest and easiest way is to fold the binding in half, wrong sides together, and press it. The edge of fabric to be neatened is then pushed in between the two layers and pinned and tacked in place.

A row of machine stitching can then be worked to hold it in place permanently – a zigzag stitch requires less precision than a straight stitch. This method does not always give a quality finish.

A better method is to apply the binding in two stages. Open out the folded edge on one side of the binding and place the crease on the stitching line of the fabric to be neatened. Pin, tack, then machine-stitch in the crease. Remove pins and tacking. Fold the binding around the edge being neatened, bringing the still-folded edge of binding to touch the row of machine stitching. This folded edge is then secured in place by hand or machine stitching. If an armhole or neck edge is being neatened, allowance must be made for the width of the binding, and the stitching line adjusted accordingly.

Interfacing is usually applied to the facing fabric to help this.

If a commercial pattern is being used to make the garment, pattern pieces for the facings will be included. Interfacing will be cut using the same pattern pieces and applied to the facing fabric. A neck edge facing may be made up of several pieces that need to be joined before the facing is attached. The 'free edge' of the facing will need to be neatened, usually by a close zigzag stitch, as this gives a flat finish. The edge of the facing is lined up with the edge to be neatened, matching dots and notches. It is pinned, tacked, then machine-stitched in place along the stitching line. Pins and tacking are removed, and the seam allowance trimmed. If the edge is curved, 'V' shapes are cut out to reduce the bulk of the seam, and the facing is turned to the inside of the garment. It is pressed in place and secured with hand stitching or understitching using the sewing machine.

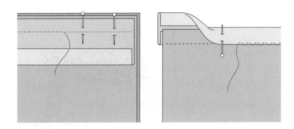

Figure 3.23 Applying bias binding

Facings

A facing is a piece of fabric cut to match the shape of the edge to be neatened. Facings are almost exclusively used on garments in places such as the fronts of shirts, blouses and dresses, armholes, neck edges, and sometimes at a waistline. As well as neatening, they increase the strength and stability of the edge.

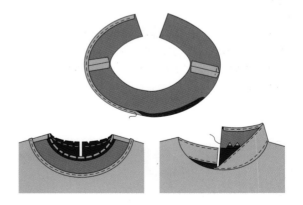

Figure 3.24 Attaching a facing

▶ Hems

Hems are found at the lower edge of a garment or sleeves. Curtains, frills and ruffles also need hems. The type of hem used depends on the fabric, the style of the garment or product, the purpose of the product and the finished effect required. The amount allowed for the hem will be stated on a commercial pattern, although this can be adjusted to make the garment fit better. The finished length required needs to be worked out, and then marked on the fabric; the hem is then worked to suit.

Narrow hem

The simplest and quickest hem is a rolled hem worked by machine. It is suitable for light- to medium-weight fabrics, but not for thick fabrics. The hem is turned up to the wrong side of the fabric and pressed to show where it needs to finish. The fabric turned in is then trimmed to an even 1 cm. The raw edge is tucked in to touch the crease, pinned, tacked and machined in place using a straight stitch. The row of stitching will be visible on the right side of the garment. Some machines have a 'hemming foot' which automatically rolls and then stitches the hem at the same time. Practice is needed when

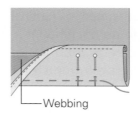

(a) Narrow machine hem

(b) Narrow hem worked with the hemming foot

Figure 3.25 A narrow machined hem and a narrow hem worked with the hemming foot.

using this attachment. Alternatively, the hem can be secured with a herringbone or slip hemming stitch so that the stitching is not visible from the right side.

Single-fold hem

This hem gives more weight to the bottom of the product or garment and is suitable for thicker fabrics as well as lighter-weight ones. More fabric is left in the hem, improving the way the fabric hangs. It also allows for the hem to be 'let down', increasing the finished length during the life of the product. This makes it suitable for children's clothes.

As before, the finished length is determined and marked, but this time the amount left for the hem is more, between 2 cm and 5 cm. The raw edge is finished, usually with a zigzag stitch, or it may be turned under and machine-stitched. The hem is then secured with a hand stitch, herringbone or slip hemming. Alternatively, the slip hemming foot could be used on a sewing machine, or a product such as Bondaweb. Bondaweb is a thin, soft, fusible fleece. It is placed between the turned-up hem and the main fabric and then pressed with a hot iron. The fleece melts and glues the hem in place. The fleece can be cut to suit the depth of the hem. This fusible fleece can also be used when working appliqué.

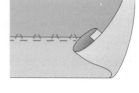

Webbing

Figure 3.26 Working a single-fold hem secured with fusible fleece and hand stitching

ACTIVITY

1. Work a sample of each of the different methods of neatening the edges of fabrics.

2. Mount each sample on a piece of paper and next to it write down how to do it. Include some diagrams.

3. Find a picture of a textile product that uses each method and mount it on the paper with the sample.

3.6 ATTACHING FASTENINGS AND OTHER COMPONENTS

LEARNING OUTCOMES

In this section you will learn about:
- attaching a range of fastenings
- attaching decorative and constructional components
- pressing and finishing.

There is a range of manufactured components available (see Chapter 2) that can be used when making textile products. They speed up the making process and improve the quality of the finished product.

EXAMINER'S TIP

In the examination you will need to be able to describe, using notes and diagrams, how to attach various fastenings and components.

❘ Attaching fastenings

Fastenings have a useful function to perform as well as sometimes being decorative. They need to be securely attached, as they will be opened and closed many times during the life of the product.

Zips

There are many different types of zips available (see Chapter 2). There are several ways of inserting a zip depending on the type of zip and the effect required. Two simple methods are described here, but there are other methods. Closed-ended zips are usually inserted into a seam, whereas open-ended zips are inserted between two sections of a product that need to be separated.

Centred zip (closed-ended zip)

This type of zip is usually inserted into a plain seam.

- Place the fabrics together, right sides facing, edges level, matching dots and balance marks. Pin and tack on the seam line.
- Machine-stitch the seam to the point where the zip will be inserted. This will be marked by a notch or a dot on a commercial pattern. Leave the tacking in place to hold the fabrics together where the zip will go.
- Neaten the edges of the seam, including where the zip will be.
- Place the zip face down on to the wrong side of the seam, placing the teeth of the zip on the centre of the seam. Pin and tack in place.
- Set up the sewing machine to do a straight stitch and change the presser foot to a zipper foot. Alter the needle position if necessary.
- Stitch round the zip, down one side, across the bottom and back up the other side, remembering to reverse at the start and end.

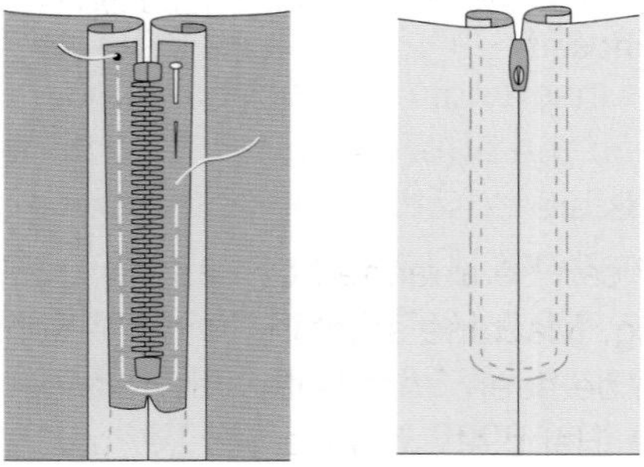

Figure 3.27 Inserting a closed-ended zip into a seam

- Remove the tacking holding the zip in place and the tacking holding the seam closed. Check the zip can be opened and closed easily. Press.

A closed-ended zip can be inserted into fabric where there is no seam by using a facing to neaten the edges of the opening made to accommodate the zip. In this case, the teeth of the zip are exposed and visible.

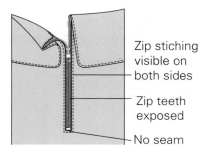

Zip stiching visible on both sides

Zip teeth exposed

No seam

Figure 3.28 Inserting a closed-ended zip without a seam

Open-ended zip

- Place the fabrics together, right sides facing, edges level, matching dots and balance marks.
- Pin and tack on the seam line but do not machine-stitch.
- Press the seam open and neaten the edges.
- Place the zip face down on to the wrong side of the seam, placing the teeth of the zip on the centre of the seam. Pin and tack in place.

- Set up the sewing machine to do a straight stitch, and change the presser foot to a zipper foot. Alter the needle position if necessary.
- Stitch down each side of the zip, reversing at the start and finish of the stitching.
- Remove the tacking holding the fabrics together. Check the zip opens and closes easily. Press.

The next part is best done with the right side facing upwards.

- Place the fastened zip right way up on a flat surface. Have the fabric the right side up, and bring the folded edges close to the teeth of the zip. Do this for both pieces of fabric, lining up the edges so that the zip is in the correct position between the fabrics. Pin and tack the fabric to the zip.
- Set up the sewing machine to do a straight stitch, and change the presser foot to a zipper foot. Alter the needle position if necessary.
- Stitch down each side of the zip, reversing at the start and finish of the stitching.
- Remove all tacking and check the zip opens and closes. Press.

Figure 3.29 Attaching an open-ended zip

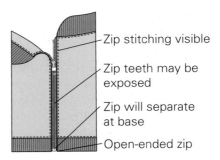

Zip stitching visible

Zip teeth may be exposed

Zip will separate at base

Open-ended zip

Figure 3.30 Inserting an open-ended zip with the teeth exposed

An open-ended zip can be inserted so that the teeth of the zip show. This method is often used when a chunky zip is used, as it forms a decorative feature. Instead of making an open seam, the seam allowance is folded in to the wrong side of the fabric, with an extra allowance for the width of the zip. The folded fabric is pinned and tacked in place.

VELCRO®

Velcro® can be attached by hand or machine stitching. Machine stitching is more secure, but can be seen from both sides of the product. Hand stitching is weaker and more time-consuming to do, but is more discreet.

In either case, the Velcro® should be pinned in place first and the positioning checked before it is permanently attached.

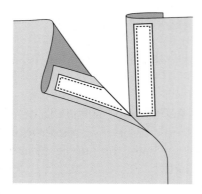

Figure 3.31 Attaching Velcro®

It is possible to buy Velcro® with an adhesive backing so that no stitching is necessary. This can be useful at times, but if the product is to be washed, or the Velcro® is in frequent use, the adhesive may weaken and the Velcro® come off.

Buttons and buttonholes

Buttons are usually hand-stitched in place, although some domestic sewing machines will attach buttons. As all sewing machines differ slightly, it is best to refer to the machine handbook for instructions as to how to do this.

A button on its own is not a fastening; it needs either a buttonhole or loop to fasten to. It is best to attach the button after the buttonhole or loop has been worked to ensure it is in the correct position.

Buttonholes

These are best worked by machine. A great deal of skill and practice is needed to make a good job of a hand-stitched buttonhole. Many sewing machines have an automatic or semi-automatic buttonhole feature, making the

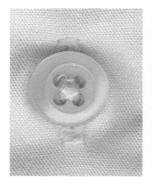
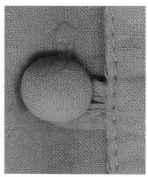

Figure 3.32 Buttonhole and loop

process quite easy. Some computerised models 'remember' the size of the buttonhole and work all the rest the same size. As all sewing machines are slightly different, it is best to refer to the handbook.

The buttonhole is worked using a close zigzag stitch and a buttonhole foot or attachment. A wide zigzag is used for the ends of the buttonhole, and a narrower one for the two sides. There is a narrow gap between the two sides of the buttonhole, and the fabric is cut between these rows after the buttonhole is stitched, usually with an unpicker. Putting a pin in at each end of the buttonhole reduces the risk of accidentally cutting through the ends of the buttonhole. The length of the buttonhole is determined by the diameter of the button, so the buttons need to be available before the buttonholes are stitched. The length of the buttonhole is the diameter of the button plus 3 mm, or if the fabric is thick, add 6 mm.

It is best to work buttonholes on double fabric, preferably with a layer of interfacing between them to strengthen and stabilise the fabric. Practise the buttonholes and test that the button fits before working them on the actual product.

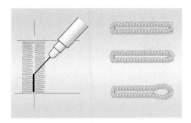

Figure 3.33 Working a buttonhole

Figure 3.34 Attaching a button with a shank

Attaching a button

There are two main types of button, those with a shank and those without. The shank creates a space between the button and the fabric it is stitched to. This allows the buttoned layer of fabric to lie flat underneath the button when the button is fastened. The thicker the fabric, the longer the shank needs to be. If the button does not have a shank, a thread one needs to be made when the button is stitched on. However, if the button is purely for decoration and will not be fastened, a shank is not needed.

- **With a shank:**
  - Position the button under the buttonhole with the wider part of the shank lying along the length of the buttonhole.
  - Thread a hand sewing needle with some machine thread that matches the fabric the button is being stitched to.
  - Sew some cast-on stitches so that the thread is attached to the fabric where the button is to be stitched.
  - Stitch through the fabric, then through the shank, keeping the stitches small and as close to the button as possible. The stitches should go right through all layers of fabric the button is being stitched to.
  - Work eight to ten stitches, then cast off.
- **Without a shank**:
  If the button does not have a shank, one

must be made from thread as the button is stitched on. This is done by placing a pin or cocktail stick on top of the button and stitching over it as the button is stitched to the fabric.

- Thread a hand sewing needle with some machine thread that matches the fabric the button is being stitched to.
- Sew some cast-on stitches so that the thread is attached to the fabric where the button is to be stitched.
- Stitch through the fabric, then through the holes in the button, keeping the stitches small and as close to the button as possible and going over the pin or cocktail stick. Alternate the holes used in the button. The stitches should go right through all layers of fabric the button is being stitched to.
- Work eight to ten stitches, then bring the needle up between the button and the fabric.
- Remove the pin or cocktail stick and pull the button to the top of the loop of thread that has been created.
- Wrap the thread round under the button, between the button and the fabric it is stitched to, four or five times, pulling it tight. This forms the shank.
- Push the needle through the fabric, close to the previous stitching, then cast off.

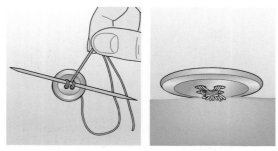

Figure 3.35 Attaching a button without a shank

Press studs

Press studs are stitched on by hand using a blanket or buttonhole stitch, which is not visible from the right side of the fabric. The two halves are lined up by putting a pin through the centre.

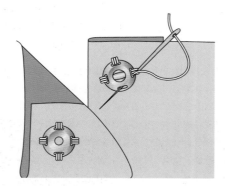

Figure 3.36 Attaching press studs

Hook and eye

A hook and eye are attached by hand stitching. The 'eye' can be a straight bar, or more of a loop, or can be substituted by a hand-worked loop. Both the hook and eye are hand-stitched in place using a buttonhole or blanket stitch, which does not go right through the fabric. Neither the hook and eye nor the stitching should be visible on the right side of the product.

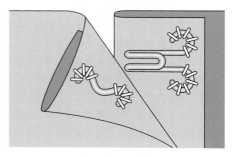

Figure 3.37 Hook and eye

Poppers

These are not stitched in place, but are sold with a tool for attaching them using a hammer. They are all slightly different and need to be attached following the instructions provided in the kit. Care needs to be taken when marking the position of the fastenings, as attaching them may create a hole in the fabric, making it impossible to reposition and reattach them if they are not in the correct place.

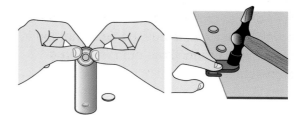

Figure 3.38 Attaching poppers

Ties

Ties can be used as fastenings. They can be made from cord, ribbon, braid, plaited yarn or fabric and can be hand-stitched in place. Machine stitching is a more secure but more visible method of attaching them. They can also be trapped in a seam. They can also be threaded through eyelets to form an adjustable fastening and decorative feature.

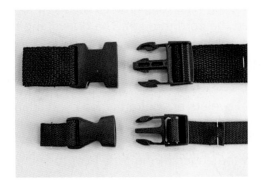

Figure 3.40 Attaching parachute clips

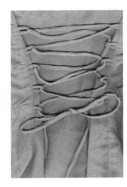

Figure 3.39 Ties used as a decorative feature, threaded through loops of fabric

Parachute clips

These are attached using braid or tape. Both parts of the parachute clip have a slot to feed the braid or tape through, which needs to be a suitable width. The tape is cut to a suitable length, then pinned on to the product, and tested to check it is in the correct position. The braid can then be machine-stitched in place, using either a straight stitch or a zigzag. It is important to ensure the end of the braid or tape has been neatened, either before use or when it is attached, so that it does not fray and allow the fastening to come loose.

Toggles

The traditional toggle fastening is usually attached using a cord. The cord is either threaded through a hole in the centre of the toggle, or fastened round the middle. The cord is then stitched on to the fabric. The toggle is fastened through a loop.

The other sort of toggle, with the locking mechanism, is used at the end of a

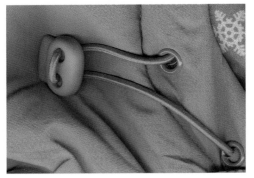

Figure 3.41 Toggles

drawstring. The toggle has a hole in it to thread the cord through. The drawstring has to be pulled up and secured in place using the toggle and released as necessary.

Eyelets and laces

Metal eyelets are not attached by stitching. A device is used to hold them in place on the fabric while they are hit with a hammer to bend the edges of the tube over to secure them in place. Sometimes they have a sharpened edge that makes the hole in the fabric. Laces can be threaded through the eyelets to act as an adjustable fastening.

Figure 3.42 An example of eyelets and laces

ACTIVITY

1. Make a collection of the different fastenings and attach them to fabric.

2. Mount each sample on a piece of paper and next to it write down how they are attached. Include some diagrams.

3. Find a picture of a textile product that uses each fastening and mount it on the paper with the sample.

Attaching decorative items

There are many decorative components that can be used to make textile products more appealing. They are not functional items; they are purely for aesthetic value.

Appliqué motifs

Appliqué is a decorative technique that involves cutting shapes out of fabric and stitching them on to another fabric. Companies produce these motifs ready to stitch on, and as well as being decorative, they can be used to strengthen and repair

Figure 3.43 An appliquéd motif

textile products. The motif needs to be pinned and tacked in position and can be machine-stitched on permanently, using a straight or zigzag machine stitch. Some have an adhesive backing. The motif is positioned on the fabric and covered with a cloth. It is then pressed with a hot iron to melt the adhesive and stick it to the fabric. It is advisable to machine-stitch the motif too in case the adhesive does not withstand wear and washing.

Ribbon

Ribbon can be stitched on to fabric by hand or machine. If a machine is used, a straight stitch, zigzag or a decorative stitch can be used. Ribbon can be attached only at one end, leaving the length free to act as ties. The ribbon can be woven through finished holes or through an open-weave fabric purely as decoration or to act as a drawstring. Ribbon can be used to form loops for buttons or to hold belts, or form straps on tops, or to weave with. As the long edges are finished, they do not need neatening, which makes it a versatile component.

Figure 3.44 Uses of ribbon

Lace

Lace can be applied in the same way as ribbon, or it can be gathered before being sewn on. It is available in a range of widths and colours. As well as being available in narrow strips like ribbon, lace motifs can be bought and stitched on to fabric. Sometimes the fabric under the lace motif can be cut away from behind the lace. If this is to be done, the motif needs to be stitched on using a close zigzag to prevent the fabric underneath from fraying when it is cut away. The same can be done with linear lace.

Lace can be used at the edge of a product, for example round cushions, at the hem of a skirt or dress, or at the bottom of sleeves.

Figure 3.45 Uses of lace

Braid

Braid can be stitched in place by hand or machine. Narrow braid may need only one row of stitching down the centre, while wider braids will need to be stitched along both edges to keep the braid flat. Braid can be used in a variety of textile products to outline designs, add a border trim or even to weave with.

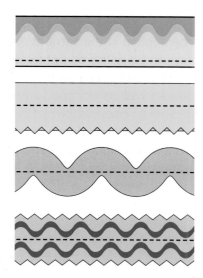

Figure 3.46 Attaching braid

Ric-rac

Ricrac can be stitched in place by machine, but the wavy construction means the stitching has to be accurately placed down the centre. It can be used in the same way as braid to outline designs and add a border.

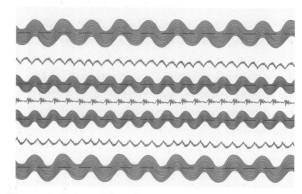

Figure 3.47 Attaching ricrac

Beads and sequins

Beads

There are many types and sizes of beads. They are very decorative, but time-consuming to apply. The position of the beads should be marked on the fabric, using a tailor's pencil, and stitched on when the item is finished. A very fine long needle, called a beading needle, and a fine thread will be needed to pass through the hole in the bead. The thread can be coated with beeswax to prevent it from twisting and breaking.

The most common beads are round rocaille beads and tubular bugle beads. In both cases the beads are stitched on by hand. The round beads are stitched with a small stitch that should be smaller than the diameter of the bead so that it cannot be seen. Alternatively, a back stitch can be used if the beads are part of a linear design. The tubular beads can also be stitched on individually or using a back stitch, or even threaded on to a length of thread and couched in place.

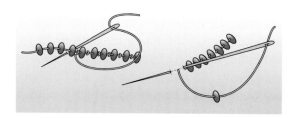

Figure 3.48 Attaching beads

Sequins

Sequins are shiny discs with a hole in the middle. They are available in a range of sizes and colours. They are attached by hand stitching when the product is finished. The position is marked with tailor's chalk on the right side of the fabric. They can be stitched on individually with three stitches going through the middle of the sequin and over the edge, or a bead can be used to hold it in the centre. They can be stitched using a back stitch if the design is a continuous line. Sequins can be bought stitched together in a strip which can be attached by couching.

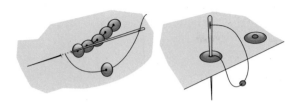

Figure 3.49 Attaching sequins

▶ Construction items

Shoulder pads

The pads are usually attached to the shoulder seam using hand stitching which is not visible from the right side of the garment. They need to be secured in more than one place so that they do not move around as the garment is being worn or washed.

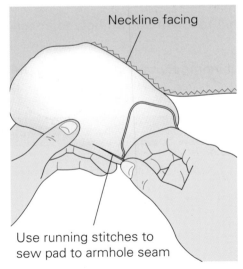

Neckline facing

Use running stitches to sew pad to armhole seam

Figure 3.50 Shoulder pads

Cuffing

The cuffing is cut to size – the finished depth of the cuff plus the seam allowance doubled. The cuffing is folded in half with the right side facing outwards. It is then slipped over the end of the sleeve, right sides together, matching the raw edges. The two are joined together using an overlocked seam, as the fabric will need to stretch.

Figure 3.51 An example of cuffing

Interfacing

Interfacing can be held in place with hand or machine stitching, or glued in place. Iron-on interfacing has a layer of adhesive on one side. The interfacing is placed in position with the adhesive next to the wrong side of the fabric and pressed with the iron. The heat from the iron melts the adhesive and it sticks the two fabrics together. While this is much quicker than hand-stitching it in place, it is a good idea to test a small sample before using it on a product in case it changes the way the fabric looks or performs.

Figure 3.52 Interfacing

ACTIVITY

1. Make a collection of the different components and attach them to fabric.

2. Mount each sample on a piece of paper and next to it write down how to attach it to fabric. Include some diagrams.

3. Find a picture of a textile product that uses each component and mount it on the paper with the sample.

3.7 PRESSING AND FINISHING

The final pressing and finishing of a textile product play an important part in achieving a good-quality product. There is a wide range of equipment available for pressing textile items (see pages 128–9). Choosing the correct equipment for the task will make the task easier and improve the final result.

Pressing

Textile products should be pressed after each stage of manufacture, after each process is completed and before starting the next. This improves the overall finish of the product and makes the final pressing easier to complete.

The final press should be about removing wrinkles, pinning and tacking marks and any shiny spots on the fabric. See pages 128–9 for information about pressing in industry.

Finishing

Finishing is another process that should be done throughout the manufacture of a product. All the raw edges of the fabrics used to make the product should have been finished as the item was being made. Loose threads should be cut off, pins removed, faults rectified and quality checks carried out.

3.8 COLOURING FABRICS

In this section you will learn about:

- methods of dyeing fabrics
- printing methods
- the advantages and disadvantages of these methods.

There are a number of different ways of adding colour to fabrics, using inks and dyes. The fabric can be immersed in a dye bath, dye can be sprayed on to a fabric, or colour can be added in the form of a paste and applied to the surface. Each method gives a different effect.

Types of dye

Cold-water dyes are effective for use on a small scale and the dye bath should be made up following the instructions provided with the dye. Salt and a fixer usually need to be added to make the dye permanent so that it does not wash out. As well as cold-water dyes, there are dyes that can be used in a microwave oven, or in a washing machine if a larger amount of fabric is being coloured. Natural fibres are the easiest to dye without specialist equipment and it is important to select the correct dye for the fibre/fabric being used.

As dyes are chemicals, so health and safety rules need to be followed carefully (see Chapter 7).

EXAMINER'S TIP

In the examination you will need to be able to describe, using notes and diagrams, how to perform the different methods of colouring fabrics.

Tie-dye

This is known as a resist method of dyeing fabric. The fabric is folded and tied tightly with string or elastic bands to prevent the dye from being absorbed evenly. Where the fabric is not folded or tied, the dye is easily absorbed giving a strong colour. In some areas no dye can be absorbed, so the fabric remains its original colour. In other areas it takes a while for the dye to soak through, resulting in a lighter colour. In this way, unique patterns can be created easily and simply.

The pattern created depends on the way the fabric is folded and where it is tied. The fabric can be knotted or twisted, stones or buttons can be tied in to create endless different patterns. More than one colour can be used on a piece of tie-dye. The fabric can be dyed and dried, then opened and refolded and tied before being dyed again.

Figure 3.53 Examples of tie-dye

Two different dye bath colours can create three different colours on the fabric.

▶ Batik

Batik is also a resist method. In this case, the absorption of the dye is prevented by applying hot wax to the fabric, using a paintbrush or a tjanting tool. Areas covered in wax cannot absorb the dye. Once the wax has been applied, the fabric is immersed in the dye bath. After the allotted time it is removed and dried. More wax can be added and then it is dyed again. This process can be repeated a number of times, to build up the design. Work through the colours from light to dark.

Interesting effects can be achieved by 'cracking' the wax, which allows the dye through to create a 'veined' effect. The wax can be scratched, or areas removed between dye baths. The dye can also be painted on to areas of the fabric rather than immersing the whole fabric, which gives more control over where the dye goes.

When the dyeing is finished, the wax is removed by ironing the fabric between two pieces of absorbent paper. When the colours have been fixed, any remaining wax is removed by placing the fabric in boiling water for a few minutes.

Another way to approach batik is to cover all of the fabric with wax, then scratch and scrape off the wax to allow the dye through. Finer, more intricate designs can be achieved using this method, known as reverse batik.

Figure 3.54 Equipment used for batik and examples of batik

Batik can be used as an industrial technique by printing gum or wax resist paste on to the fabric using hot rollers. Alternatively, resin can be printed on to the fabric to resist the dye.

The fabric is then dyed, and the wax or gum removed using heat, which also fixes the dye.

Diffusing

Dye can be sprayed on to fabric using a spray diffuser. The dye needs to be thin, like water. One end of the diffuser is placed in the dye, and the other end placed in the mouth. When you blow down the diffuser, a fine mist of spray comes out of the other end and is directed at the fabric. The idea is to create a lightly speckled effect on the fabric.

If the diffuser is too close to the fabric, or too much dye is applied, the dye will soak into the fabric and spread rather than sitting on the surface, which will spoil the effect. Ink can be used instead of dye, but this may not be colour-fast.

The application of the dye can be controlled using a stencil. A stencil is a piece of paper with a hole cut in it. The paper protects the fabric from the dye, only allowing it through the hole. Alternatively, several colours can be sprayed over each other and blended to create an interesting effect.

An airbrush can be used to cover larger areas of fabric. This uses a compressor to blow the dye on to the fabric, giving a more even coating and covering large areas quickly. This process can be done on a large scale for industrial production.

Block printing

This is a traditional method of printing. A 'block' is made from a resistant material, usually wood but sometimes rubber or metal. The design is marked on to the surface of the block and the background is cut away, leaving the design raised. This used to be done by hand, but now can be done by a milling machine or laser cutter.

Dye is applied to the surface of the block, using a roller, or the block is dipped into the dye, then the block is pressed on the fabric. The blocks need to be carefully lined up to create a repeat pattern, and the way they are lined up can be varied to create different effects. A number of blocks can be used on one fabric to build up a more complex design, and different colours can be applied using different blocks.

Block printing is labour-intensive and therefore expensive. This technique has been developed into roller printing for industrial production (see Chapter 2).

Figure 3.55 A spray diffuser and an example of diffusing

Figure 3.56 Block printing

ACTIVITY

1. Produce a sample of each method of colouring fabric.

2. Mount each sample on a piece of paper and next to it write down how it was produced. Include some diagrams.

3. Find a picture of a textile product that is decorated using each technique and mount it on the paper with the sample.

3.9 DECORATIVE TECHNIQUES USING FABRIC AND THREAD

LEARNING OUTCOMES

In this section you will learn about:
- methods of decorating fabric with fabric and thread
- the effect produced by these methods.

As well as dyeing and printing on fabrics to make them more attractive, designs can be added using fabric and thread. Several of these techniques are good examples of the 6Rs, including recycle, rethink, repair, reuse and reduce.

EXAMINER'S TIP

In the examination you will need to be able to describe, using notes and diagrams, how to work the different methods of decorating fabrics.

Quilting

There are several types of quilting. English quilting involves stitching three layers of fabric together. A layer of padding, such as polyester wadding, is placed between a top layer, usually a decorative fabric, and a bottom layer, which is usually a cheaper, plainer fabric. The two outer layers of fabrics have their right sides facing outwards. If the quilted fabric is to be reversible, a good-quality fabric will be used for both outer layers.

The layers are placed together and pinned, starting in the middle and working outwards. The layers are then tacked, starting in the centre and working outwards in a spiral. A frame can be used to hold the layers together and keep them flat.

The machine stitching used to hold the layers together should be placed close enough to hold the layers securely but not so close as to flatten the wadding and squash out all of the air. A longer machine stitch is used to accommodate the thickness of the fabric. Interesting patterns can be created with the stitching, the placement and the type of stitch used. The stitching lines can be marked using tailor's chalk or pencil, or a quilting foot can be used on the sewing machine. Some machines have a 'walking foot' which releases the pressure on the fabric as it moves through the machine, making the stitching easier and improving the quality of the finished result.

This type of quilted fabric is not only decorative, but is also a good insulator and has protective qualities. It is advisable to quilt the fabric before cutting the pieces out to make a product, as when fabrics are quilted, the overall size may be reduced.

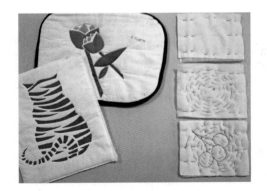

Figure 3.57 Examples of quilting

Trapunto

Trapunto, or padded quilting, is another form of quilting. Two layers of fabric are placed together, right sides facing out, then pinned and tacked. Shapes are marked out on the top fabric and stitched round using a sewing machine. A small cut is made in the backing fabric within the shape, and the shape is padded with a filling such as polyester fibres. The slit is then sewn up by hand.

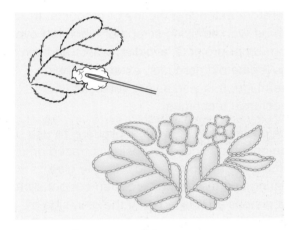

Figure 3.58 Trapunto quilting

Italian quilting

Italian quilting, also known as corded quilting, gives a linear effect. Two fabrics are placed right sides facing outwards, then pinned and tacked together. Parallel lines of machine stitching are worked on the fabric, and cord threaded between them, working from the wrong side.

Figure 3.59 Italian quilting

Appliqué

Appliqué involves cutting shapes out of fabric and stitching them on to a backing fabric. Appliqué is an excellent technique for repairing and strengthening fabrics and therefore extending the life of products. It can be worked using scraps of fabric left over from other projects, avoiding waste and can rejuvenate old products, extending their useful life. It is also a very versatile, decorative technique.

The easiest way to work appliqué is to use iron-on interfacing. The design is traced on to the side of the interfacing with the glue coating and the shapes are then cut out. The cut shapes are placed with the glue side of the interfacing next to the wrong side of the fabric to be applied. The shapes are then ironed to make them stick to the fabric, and then cut round again. The interfacing prevents the edges of the fabric fraying or curling and stiffens the fabric slightly, making it easier to handle. It also cuts out the need to make a paper pattern or template. The shapes are then pinned and tacked in place, before being machined using a zigzag stitch worked over the edge of the shape. Bondaweb can be used to hold the shapes in place rather than pinning and tacking.

There are alternative methods that can be used. A paper pattern can be made for each of the shapes in the design, with a seam allowance added which is turned under when the shape is stitched on to the background fabric. The shapes can be hand-stitched using blanket stitch or herringbone stitch for a decorative finish, or if the appliqué is being worked on a product which can't be positioned under the machine needle.

Figure 3.60 Examples of appliqué and padded appliqué

Appliqué shapes can be padded in the same way as trapunto quilting. Another variation is reverse appliqué, sometimes known as San Blas appliqué, or mola work. Several layers of fabric are placed on top of each other, and pinned and tacked in place. The shapes of the design are marked out on the top fabric using tailor's pencil and then stitched round using the sewing machine. If a zigzag stitch is used, it will prevent the fabrics from fraying. The top layers are cut away to reveal those beneath.

Figure 3.61 Example of San Blas appliqué

Free machine embroidery

This technique can be used to repair holes in fabric and strengthen the fabric where it has worn thin. It has been developed from a darning technique worked using a sewing machine. The machine is threaded as usual, the presser foot is either removed or replaced with the darning foot and the feed dogs are disengaged. The fabric is stretched tightly in an embroidery ring, which is used upside down so that the fabric sits flat on the bed of the machine. The design to be stitched is marked out on the fabric. Large areas of the design can be filled in using a large zigzag stitch, while fine lines can be worked using a straight stitch. The fabric is placed under the machine needle and the presser foot lowered to ensure correct tension. The fabric is moved around under the needle to fill in the design with stitching.

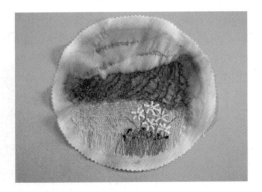

Figure 3.62 Example of free machine embroidery

Computer-controlled embroidery

Computer-controlled sewing machines range from the relatively simple to the very complex. Some of these machines will work in the same way as a conventional sewing machine, producing rows of stitching, and can be used in the same way as a conventional machine for general sewing as well as for decorative stitching.

Others are dedicated to producing decorative stitching in the form of a motif. They do not have feed dogs and use a presser foot like a darning foot. They usually have some built-in stitch designs and many allow some editing of these. They often have some text facility to allow lettering to be stitched, with some choice of font and size. More expensive models are more versatile and allow for more development of the built-in stitch patterns.

These machines work following the principles of free machine embroidery. The machine has a moving arm with a hoop attached. A backing fabric or stabilising fabric, rather like iron-on interfacing, is ironed on to the reverse of the fabric, which is then put into the hoop. The machine is set up and programmed following the instructions in the handbook. The arm moves the hoop around under the needle to stitch the design. It will stop and beep when the thread needs to be changed to a different colour, or if there is a problem such as the thread breaking or running out.

Some of these machines allow a design to be created in a drawing package, or using Clip Art images, which can then be converted to a stitch design using software provided with the machine. A memory device is used to transfer the information from the computer to the sewing machine, or the machine can be connected directly to the computer. In this way, unique individual designs can be stitched.

Some machines can be connected to the internet and stitch patterns downloaded to the machine, or additional stitch patterns can be purchased on a memory device that is put into the machine and the information downloaded.

Figure 3.63 Computer-controlled embroidery

Patchwork

Patchwork is the joining together of smaller pieces of fabric to create a larger piece. It is a very old technique dating back to the eighteenth century and was traditionally used as a functional technique for bedding, both in England and by colonists in America. Patchwork was developed as a way of using every scrap of fabric, recycling and reusing fabrics to prevent waste.

There are many different designs and styles of patchwork, but they fall into two main categories. The simplest is one-shape patchwork in which all the pieces are the same size and shape. The other is block-unit patchwork, where the pieces are joined into a block – often a square – which then becomes the basic shape. These blocks are then joined together. Crazy patchwork is a form of block patchwork where random shapes are used to create the block, and different shapes are

used for each block. The effectiveness of patchwork relies on the careful choice of fabrics and colours.

The patchwork shapes can be joined by hand stitching or using the sewing machine. A template or pattern is needed to ensure the shapes are cut accurately, or they will not join up properly. Traditionally, a paper pattern was made for each shape, without a seam allowance. The paper pattern was then pinned on to the wrong side of the fabric and cut out, leaving a small seam allowance. The seam allowance was then folded in and tacked in place through the paper pattern. This was done for all of the pieces needed. To join the pieces, they were placed right sides together and joined using an overcasting stitch. Another way to join the pieces would be to place pieces side by side with the edges together, then use a wide zigzag stitch on the machine to join them.

Another way to work is to make a paper pattern with a set seam allowance, which can be used to cut the pieces out but without being attached to them. Once cut out, the pieces can be joined together using an open seam. This works best when the edges of the shapes are straight and the pattern is regular.

For a creative piece of patchwork where various shapes are used, like the duvet cover shown in Figure 3.64, iron-on interfacing can be used. The design is drawn out full size, and the interfacing placed on top with the glue side facing up. The design is traced on to the interfacing, then all the pieces cut out. A numbering system may be needed to identify the pieces if the design is complex.

The interfacing is then ironed on to the reverse of the fabric being used, and a seam allowance added when cutting out.

The pieces can be joined by hand or machine.

In the example shown, the pieces were joined by hand, the other parts of the design joined by machine.

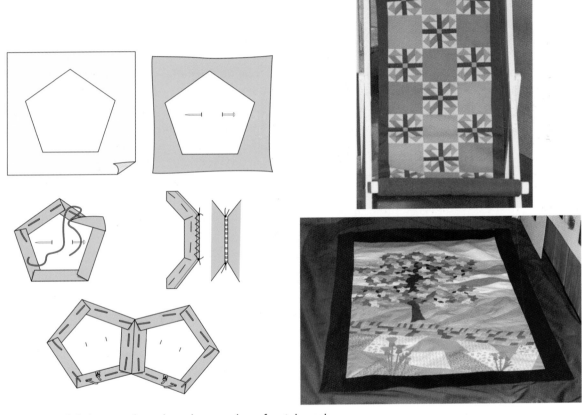

Figure 3.64 Joining patchwork and examples of patchwork

ACTIVITY

1. Produce a sample of each method of decorating fabric, using fabric and thread.

2. Mount each sample on a piece of paper and next to it write down how to do it. Include some diagrams.

3. Find a picture of a textile product that is decorated using each technique and mount it on the paper with the sample.

CHAPTER 4

COMPUTER APPLICATIONS

**By the end of this section you should have developed
a knowledge and understanding of:**

- The use of CAD in the school environment for producing drawings and 2-D and 3-D images
- Onscreen modelling and manipulation of images
- Appropriate use of text, database and graphics software in school and commercial situations
- Storing and sharing data electronically
- The application of CAD/CAM/CIM to the designing and making of models and prototypes
- The application of CAD/CAM to one-off and quantity production
- Computer control of machines (CNC) including sewing and embroidery machines, fabric printing, fabric finishing and cutters

Vast improvements and advances in technology, the quality, user friendliness and price of computer systems and software for the textile sector have gradually changed the way designing and manufacturing within this industry takes place. You will need to have a clear understanding of the different ways computers can be used to enhance your own methods of designing and making a textile product, as well as having knowledge of how computer applications are used within industry.

4.1 USE OF ICT/CAD TO GENERATE DESIGN IDEAS

CAD software can be used to produce accurate drawings of a textile product and its components. Fashion CAD enables the user to design prototype garments in a 2-D or 3-D format, which allows all viewpoints to be looked at, using a rotate facility. Areas of the design can also be enlarged and amended onscreen. As a learner, it is important that you are creative in the way that you incorporate ICT into your design work, using appropriate techniques to give the best end result. There are a variety of ways you can do this.

Onscreen modelling and manipulation of images

You do not have to be able to use a design or draw package to be able to create interesting effects using ICT. Try this simple method of free illustration and scanning to achieve something different (as illustrated in Figure 3.1).

1. The completed hand-sketched garment design is scanned and saved as a JPEG image.

2. Insert the JPEG image of your design on to a Microsoft® Office Publisher page. Using the picture toolbar, select the pencil icon, set transparent colour (second from the right) and right click on the background of your design.

Figure 4.2 The picture toolbar from Microsoft® Office Publisher

This tool will remove areas of your design to produce the effect shown on this illustration. Experiment with this technique until you achieve an outcome you are pleased with. Drag the picture off the document page and save the picture as a JPEG image.

3. Print a copy of your modified image on to A4 paper and enhance, using a black fine-liner pen to highlight important details.

There are many more ways you can design using a scanner. Try some of the following:

- Scan in a template of a garment or textile product you want to work on. Flood-fill the outline to create different colours and patterns.

- Scan samples of fabrics, threads, yarns and components directly on to your design page. These could be worked pieces of

Figure 4.1 On screen modelling

Figure 4.3 Fabrics can be scanned into the computer and manipulated to create realistic 2-D designs

embroidery, embellished cloth or tie-dye pieces which can then be cropped into garment shapes onscreen.

- Scan your own logos, motifs or pattern ideas on to your design pages. Backgrounds for designs can be created in this way to drop your illustrations on to. Try scanning actual objects, like leaves and feathers, to show where you have gained your inspiration or explain a starting point.

- Use the scanner to produce designs to be printed on to iron-on transfer papers which can be translated straight on to fabric. Remember to 'flip' the image before printing so the correct view is shown on the finished piece!

Graphics programs and design software

There are many ways that ICT can be used to design and model your ideas in the classroom. Of course, this depends upon what is available to you and how confident you are in using design programs. ICT should be used to enhance your controlled assessment units and will help you to present, communicate, generate and model your thoughts throughout the portfolio. Here are a few helpful hints to get you started:

- Try using the many graphic programs to model new styles and ideas and to create packaging and care labels, for example Microsoft® Word, Paint and Office Publisher.

- Experiment with pattern tessellation and repeats. Templates for techniques such as patchwork and quilting can be manipulated to achieve surface patterns. TechSoft 2D Design allows you to create these effects easily with the mouse.

- Texture mapping a garment with pattern and colour ideas can be created using Paint, CorelDRAW®, Paint Shop Pro® or Adobe® Photoshop®.

- Line drawings of ideas or templates of basic garment shapes can be grouped on an image board by cutting and pasting originals, saving as JPEG images, and dropping on to a prepared background.

Figure 4.4 Vector-based illustration programs, like Adobe® Illustrator®, are used in industry for drawing lines, curves and geometric shapes

- Annotate your designs using Microsoft® Word with different font styles related to different themes.

- Adobe® Photoshop® can be used to create a variety of styles using specific filters to create mood, texture and pattern. Interesting filters to experiment with include texture effects: soft plastic, rough leather and tiles. 3-D effects and artistic effects also have some great background possibilities for design work.

- The use of digital photography is very important to show how your product fits, especially at the modelling stage. Photographs of different stages of development can easily be annotated onscreen to highlight potential areas for development. Different colourways can also be modelled using the digital images in Adobe® Photoshop®.

- Design database software provides ready-made images that can be modified and manipulated.

- Use of a graphics tablet to create or reproduce drawings is useful to trace existing body stencils, garment outlines and photographs. The tablet also allows you to draw freehand to achieve more accurate and personal ideas.

In industry, there are two ways computers can deal with visual information: vectors and bitmaps.

Vectors

This type of design program is used to draw lines, curves and geometric shapes because it enables the designer to produce smooth lines. Vector files are economical on memory and patterns can be transmitted around the world quickly and accurately. Vector-based programs such as Adobe® Illustrator®, CorelDRAW® and Macromedia® FreeHand® also allow the designer to colour-fill, use text and text-wrapping, produce seamless pattern repeats and fills and add numerous filters. Hand-drawn illustrations can be scanned into the program and converted easily into a vector drawing.

Bitmap

This type of program is best suited to real images such as photographs and is based on a collection of pixels. These pixels can be edited and adjusted to give more detail to a design. Small and fairly flat objects like buttons, trimmings and yarn can be scanned into the computer, saved as a bitmap image and then combined with the design. Creative backgrounds can also be manipulated and developed using existing products like magazine pictures, food wrappers, etc. which can be tidied up onscreen.

 4.2 DATABASES AND SPREADSHEETS – STORING AND SHARING DATA

A database program is primarily used to store and organise information (data) and is useful as a research tool. Data can be edited and retrieved using keywords which are categorised under specific headings, for example, fabric features, characteristics and properties. The program is basically a table made up of rows and columns (fields) which you can add to and extend, depending on the nature of your data.

Databases and spreadsheets can also be used to model costs and production processes. A spreadsheet program such as

Microsoft® Excel is useful for interpreting, calculating and presenting information for your controlled assessment units, particularly data relevant to testing. A big advantage of using electronic spreadsheets is the ability to link data so that all related figures are automatically updated.

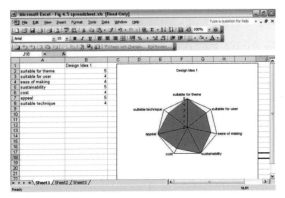

Figure 4.5 Using a spreadsheet to show the evaluation of a design idea

KEY POINTS

- ICT is a valuable tool for enhancing all aspects of your controlled assessment portfolio.
- ICT saves time, enables accuracy and allows you to be creative when designing and making a textile product.

ACTIVITY

Create a fabric database – collate and scan as many fabric types as you can. Classify the fabrics under specific headings of your choice, for example, natural, modern, woven, knitted, etc. You may want to extend your database to include care instructions, suitable products and price per metre. Present your database to your textile group.

4.3 THE APPLICATION OF CAD/CAM/CIM IN THE DESIGNING AND MAKING OF PROTOTYPES

You will need to have an understanding of how CAD, CAM and CIM are used in the designing and making of prototypes and models both in the classroom and in industry. CAD (computer-aided design) and CAM (computer-aided manufacture) are popular terms used frequently in the classroom. Less recognised is CIM (computer-integrated manufacturing) which incorporates the whole of the production process, from product concept through to manufacture and despatch, using computers in an integrated and efficient way.

Modelling of prototypes

Modelling involves the designing, developing and modification of ideas to find out what works best. This can be done either by using a computer or by making a model by hand.

Modelling in industry	Modelling in the classroom
Within industry, computer-aided design and automation systems allow the designer to create virtual prototypes of garments, with realistic fabric behaviour (i.e. drape) from designs: this is referred to as texture mapping. This process involves the transfer of a 2-D flat pattern to a 3-D virtual dress form which then creates a 3-D computer-modelled prototype garment. The system includes a 3-D body model and a virtual prototype model. A grid is drawn or printed on a fabric and the fabric is draped on the 3-D body. Photographs are taken of the garment and entered into the system. A picture converter shows views of the garment from different angles and stores the images in a picture base. This allows for more accurate and consistent texture mapping of surface and construction details on to the garment. The majority of designers, however, still prefer to complete a prototype by hand, using a cheap fabric (calico). Once the prototypes are completed, initial costings for material, labour and profit margins are calculated and compared with street prices. Those selected are then tested for sales appeal at exhibitions, ready-to-wear shows or through one-to-one appointments. Styles which lack appeal are reworked to meet the needs of the consumer.	For your controlled assessment portfolio A571, you will be required to evaluate the processes involved in making the prototype. It is important, therefore, that you are able to show all developmental work, preferably through modelling, to be able to attain the higher marks. This can be achieved in a number of ways: • Making small samples of important details, e.g. cuff detail, collar shape and construction details (seams, shaping and reducing fullness, etc.) • Creating a toile or mock-up of the textile product. This can be completed in paper or fabric (cotton or cotton calico). • Investigation into techniques, e.g. different types of worked embroidery, applying colour, etc. • Manipulating and selecting fabrics, comparing the properties to ensure suitability and fitness for purpose. Completing a toile will help you to find out what your 2-D designs look like in 3-D. This will help you to make informed decisions and evaluation points about modifications needed to the final product. Digital photography is useful at the modelling stage to record key areas of the making process. These images can be manipulated and annotated to show developments and modifications resulting from sustained and relevant investigation.

Table 4.1 Modelling

Figure 4.6 CAD systems allow designers to create virtual prototypes of garments

Figure 4.7 A toile gives a realistic idea of how the final design proposal will take shape

4.4 THE APPLICATION OF CAD/CAM TO ONE-OFF AND QUANTITY PRODUCTION

Many CAM machines are linked directly to a CAD program, which means that designs can be fed directly to the machines to complete the task. Once designing and modelling have taken place, the item needs to be manufactured. In industry, there is a wide range of automated industrial machinery to deal with any technique and construction method required to produce a quality product. Within the classroom, the opportunities you have to use CAM may be limited. However, awareness and understanding of what is available can help you to structure your portfolio and ideas, making reference to what could improve your manufacturing strategy if the machines were available.

Sewing machines in industry

Industrial sewing machines are built to withstand rigorous and continuous use at high sewing speeds, reducing the amount of time taken to complete a textile item. Automated sewing systems or CNC (computerised numerical control) machines are controlled through a computer, where the sewing cycle is stored as a program. The machines can perform specialised functions to suit any construction process from buttonholes, sewing on the buttons and overlocking, to embroidery and printing.

Computerised embroidery machines

With advancements in technology, stitching and embroidery machines are constantly changing to satisfy the demands of the consumer. Machines need to be fast and easy to use, offering a range of facilities to suit any textile product. In the classroom, the same ease of use and range of facilities are required from a sewing machine in order to achieve a quality end product.

Computerised embroidery machines allow the user to create personal designs, which can be modified on the computer or on the sewing machine's built-in computer screen and transferred into stitch format through

Computerised embroidery in industry	Computerised embroidery in the classroom
Embroidery is usually worked using tubular system machines which are often equipped with special attachments for cord/loop, boring and/or sequin embroidery. Figure 4.8 shows a multi-head bridge-type embroidery machine with a drop-table for embroidery on caps/hats, made-up garments, other tubular goods and conventional panel/border embroidery. When changing from standard embroidery to caps for example, the machine is equipped with a special quick-change needle plate insert which can be changed in seconds. These machines are accurate and efficient once programmed.	Computerised embroidery machines can be programmed to perform a variety of functions: • Different stitching functions, from straight lines to pattern stitching, which can be programmed into the machine's memory. • Repetitive construction techniques, such as working identical buttonholes. • Pre-programmed motif and logo designs, which can be stitched from the sewing machine's memory bank or database. • Free illustration ideas which can be scanned and stored on a memory card, then transferred to the sewing machine.

Table 4.2 Computerised embroidery

Figure 4.8 A multi-head tubular system embroidery machine

customising and digitising systems. Designs can be combined with cut, copy and paste facilities. Images can also be created from Clip Art files, personal drawings and photographs. It is very quick and easy to experiment and explore a range of surface effects and pattern styles using a computerised embroidery machine. New 4-D software systems have been developed to add extra facilities to help you create personal embroidery designs. The user is now able to create, edit and display all designs in true 3-D reality with zoom ability as well as being able to choose background colours and select from different fabric textures.

Computerised knitting machines

Knitting technology today can produce fabrics and garments at high speed that are creative

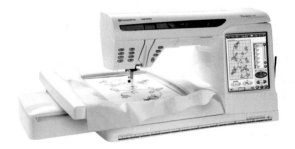

Figure 4.9 Using a computerised embroidery machine to create a quality product in the classroom

and complex. Garment designs can be knitted as individual pieces (cut-and-sew) or knitted as a whole item without seams (3-D knitting), which is much less wasteful and far quicker. This type of knitting facility can ensure that single and customised items can be made quickly to order, alongside batches. The use of CAD also allows knitting designs to be changed quickly and respond to fashion trends. Examples of electronic machines are Stoll, Shima Seiki and Protti.

Computer-controlled weaving

Computer-aided weave can produce fabrics with many layers and surfaces. Designs can be transferred straight to the weaving loom and then woven using computer-controlled looms. Type of weave, size and colour of the resulting fabric can be controlled and changed; for example, in jean production sometimes the denim is woven in a narrow strip so that the selvedge can be a feature in the trouser leg.

Lay plans and cutting out

In industry pattern lays are designed using CAD and then transferred to the CAM machines. Chapter 5 gives more information about industrial lay plans and cutting technology. Once the pattern lay is finalised and stored, the cutting instructions are sent directly to the cutting machine. This increases the efficiency of the process because it is faster, more accurate and less costly than one person cutting out!

Digital printing

Digital printing allows the designer to work straight from the computer to the fabric,

Figure 4.10 The laser cutter can cut many layers of fabric at once

leaving out the paper design process. Very high-definition imaging can be achieved and many colours can be printed with the use of very few screens. There is no limit to the number of layers that can be used within a design, repeats can be any size and 3-D imagery can also be achieved. The use of pigment-based inks instead of dyes has meant that the process is more efficient, environmentally friendly and economical in terms of running costs. Pigments also adhere better to a wider range of fabrics, giving a better quality finish.

Figure 4.11 Digital printing of textiles

KEY POINTS

- CAD/CAM/CIM can be used in one-off and quantity production.
- Computer-controlled machines (CNC) can increase efficiency and accuracy in sewing, embroidery, printing, finishing and cutting processes within industry.

ACTIVITY

1. Using the internet as a research tool, collect information about the different types of embroidery machines available on the market, for both commercial and industrial use. Present your findings.

2. Using the information you have gathered for Question 1 to help you, explain why it would be more economical to use multi-head computerised embroidery technology than a commercial embroidery machine.

CHAPTER 5

INDUSTRIAL PRODUCTION

By the end of this section you should have developed a knowledge and understanding of:

- The appropriate application of one-off, batch and just-in-time (JIT) production methods to the manufacture of textile products
- Methods of pattern production and grading
- Methods of producing pattern lays
- Methods of laying out fabric and cutting out
- Finishing
- Labelling and tagging of textile products, including smart labelling and radio frequency identification (RFID) tagging
- Health and safety

You will need to have a clear understanding of the different types of production methods that are appropriate to the textile products you design and eventually make.

5.1 BASIC COMMERCIAL PRODUCTION METHODS

Two of the most common methods of commercial production you will need to know about are one-off production and batch production.

One-off production

One-off is a traditional method of production where one unique item is made in response to a brief or a specific user request. This method may also be referred to as individual, job or bespoke production.

What does one-off production involve?

- Highly skilled operators, where one operator or a team makes the whole of the textile product from concept to completion.

- Work is usually detailed and extensive and often hand-pieced.
- Versatile machinery is used which can be adapted to suit any textile process.
- It is labour-intensive.
- The process involves high throughput time (i.e. takes a long time to make a textile item in comparison with batch production).
- Usually made from expensive fabrics and manufactured components.
- End product is of a high quality and cost.

Figure 5.1 A single, unique garment made to the specification requirements of a single user is an example of bespoke clothing

Batch production

Batch production is a method used to produce a specific number of identical items at the same time. This method provides small quantities of textile products for order, to use for a specific user or purpose.

What does batch production involve?

- It is flexible and easily changed to meet the demands of the target market.
- Batches can be repeated as many times as required.

- Production costs are relatively inexpensive.
- A variety of textile items can be manufactured, although time can be lost during the resetting of machinery for each new product.
- Workers can specialise along a production run: for example, sewing a cuff on to a sleeve. However, this has the disadvantage of making their job become repetitive and boring.
- Workers have access to more flexible working conditions and training needs.
- Stock may need to be stored before despatch. If products remain unsold this leads to wastage.

Just-in-time

Another commercial manufacturing system you will need to know about is **just-in-time** (JIT). This type of production helps the manufacturer to manage and control stock. Regular deliveries of materials and components from suppliers arrive at the manufacturer 'just in time' for production and are used immediately. The advantages of this type of production method are many, from lower production costs through to fast, efficient and high-quality performance. Problems however, can occur if deliveries of materials are delayed, bringing production to a halt.

KEY POINT

- It is important that manufacturers select the appropriate production method for the textile products they make.

5.2 METHODS OF PATTERN PRODUCTION AND GRADING

You will need to understand about the different types of patterns and how they are used both in industry and in the classroom when designing and making a textile item. Creating a pattern is an important part of the construction process and must be completed correctly to ensure that the end product is of a good quality and fit for purpose.

The pattern process

In the clothing industry patterns are developed from basic blocks. These are two-dimensional shapes made from a standard set of measurements to suit a range of body sizes. The blocks have to conform to the British Standards Institute (BSI) standard sizing. This information, gathered from regular market surveys, is very important to clothing manufacturers as it enables them to ensure a good fit for a range of sizes.

A basic block pattern is made for each part of a garment which when fitted together produces a basic garment shape. The block can be used over and over again, so it is usually made from heavy cardboard or plastic to withstand lots of handling.

A pattern drafter uses the blocks to further develop the basic garment shape by adding style lines, construction details and decorative features to create a new pattern. A garment toile is then produced from the new pattern so that the basic fit or 'ease' can be adjusted if required. One-off or bespoke garments have blocks drafted from the personal measurements of the client from which all their clothes can be made.

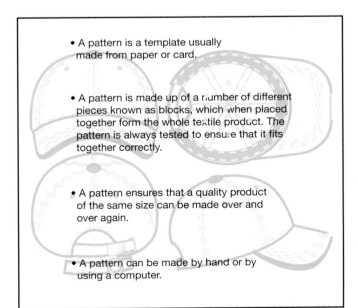

• A pattern is a template usually made from paper or card.

• A pattern is made up of a number of different pieces known as blocks, which when placed together form the whole textile product. The pattern is always tested to ensure that it fits together correctly.

• A pattern ensures that a quality product of the same size can be made over and over again.

• A pattern can be made by hand or by using a computer.

Figure 5.2 Why is a pattern useful?

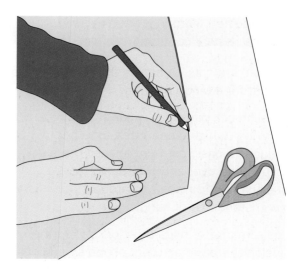

Figure 5.3 Construction of a garment using a basic block pattern as a starting point

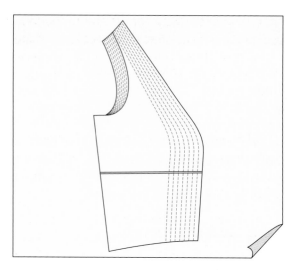

Figure 5.5 Finished pattern piece showing grading lines

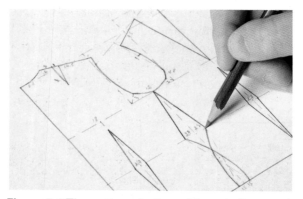

Figure 5.4 The pattern drafter adds style lines and construction details to the block to give shape to a garment

Pattern grading

Grading is the process of taking a finished pattern and increasing or decreasing it to make different sizes. Once a pattern has been made, approved and graded properly, all the design lines are proportioned throughout the size range. You can see this process clearly on a multi-size commercial pattern.

In industry, grading is done using a CAD system with a patternmaker. This guides a cursor around the edges of the pattern and automatically calculates the grading, adding it to the pattern piece.

Pattern markings

When the pattern has been designed and graded it is ready to be marked up using symbols. These help you to make and fit the garment successfully. Always read all the instructions carefully before starting to use the pattern. More information relating to pattern symbols can be found in Chapter 3.

Commercial patterns

A commercial pattern is a ready-made set of paper templates which can be bought from shops and is usually mass produced from thin tissue packaged in envelopes. The actual production of a commercial pattern is not time consuming, nor expensive. It is the design of the pattern that is the most time-consuming and costly part of its production. A successful pattern enables the user to make an article of clothing for a fraction of the cost it would take to buy a garment ready-made in a store.

How a commercial pattern is made

1. Preliminary sketches from the designer are discussed by marketers, dress designers, dressmakers and the design department.

2. Members of the technical department (design merchandising, product standards, patternmaking and dressmaking) decide details of a style and determine the construction details.

3. Decisions are made about the number of pattern pieces, the style number based on degree of difficulty, suitable fabrics, sizes the patterns will be graded to, and how it will be made.

4. The patternmaker creates the first pattern. The paper pattern is drafted on to muslin (a plain fabric) which is used to make up a 'drape' or sample garment. The drape is pinned in place and hand-stitched and is then thoroughly reviewed by both the patternmaker and the designer. Adjustments are made where needed.

5. When the drape is approved, the pattern is sent to the computer-aided design (CAD/CAM) department. The technician digitises the basic pattern pieces. All the separate pattern pieces are then blocked, which means they are marked with all the important information (seam allowances, fold lines, dart lines, etc.) needed to make them usable pattern pieces. After blocking, the pieces are plotted using a laser plotter.

6. The pattern is then graded to the various sizes using a computer program (see Figure 5.5).

7. The measuring department works out the fabric yardage and **notions** needed. Computer software helps the technicians create the best fabric layout so that the fabric can be used efficiently. Step-by-step instructions are written for the consumer in easy-to-understand language.

8. A computer template (or plot) is used to plot out the pattern. The pattern pieces are then laid out, printed and folded to fit into the envelope with the instruction sheets.

A selection of the completed patterns is carefully tested using a small production run of selling samples, and the style is presented to buyers in wholesale markets. Once the style has demonstrated sales potential, the pattern is graded for a range of sizes, vetted and approved for production.

Figure 5.6 The drape is a sample piece made so the pattern can be reviewed

KEY POINT

- In industry a pattern is the most important element in the development of a garment. If the pattern does not fit correctly, the design will not look right.

ACTIVITY

Discuss the quality-control points that might have taken place before a commercial pattern reaches its point of sale.

5.3 PRODUCING LAY PLANS

As a learner you will need to know about lay planning and the effective use of fabric to avoid waste. You will also need to have an awareness of how CAD and CAM are used to produce a pattern lay in industry.

What is a pattern lay?

A pattern lay is an arrangement of pattern pieces on a width of fabric. The pattern is arranged on the fabric, following the instructions and symbols printed on the pattern piece. For example, the 'straight grain arrow' signals the direction the pattern should be placed on the fabric.

In the classroom, you can produce a pattern lay by simply arranging the pattern pieces on your fabric manually. Move the pattern pieces into different positions until you find the best fit with the least waste. This method is suitable for a one-off item, but is not practical for a large batch of textile products.

Pattern lays in industry

Computers are extensively used for pattern lays in the fashion industry. The process used is referred to as computerised marker making. The purposes of marker making are:

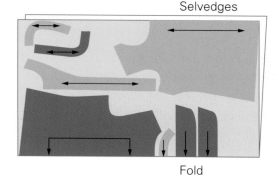

Figure 5.7 Pattern layout for a fabric 150 cm wide

- to make a layout for the cutter to follow
- to place the pattern pieces close together to avoid fabric waste
- to plan and organise the cutting order to ensure that the correct quantities of each size are cut out.

Various methods can be used to produce a pattern lay; the most commonly used is where patterns are spread out so that each size and colour is cut for a number of garments all at once, saving time and fabric. It is also possible that the grain directions, fabric types, one-way points, plaids, stripes and naps, etc. can be considered while

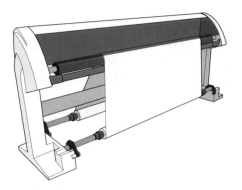

Figure 5.8 Pattern marker in use

making the marker. Miniature images of the graded pattern pieces are displayed graphically on the computer screen during this process.

With the help of the computer and the display, the operator can position the pattern pieces into the most **efficient** layout plan.

Once the marker is completed, the lay plan is printed on to a long sheet of paper with the help of a plotter. The arrangement of pattern pieces is then checked, both on the screen and on the miniature printout drawn from the plotter. The layout can then be modified and/or saved for future use.

Fabric is rarely folded in industry and pattern lays are usually arranged on single, flat fabric. The pattern pieces therefore are made for front and back and left and right sides of a garment. In the classroom, you will fold the fabric you are using in half to create two layers and place your pattern pieces on the folded fabric. This would mean that both sides of the garment can be cut out at the same time.

In industry the cutting instructions are sent directly to the cutting machine once the lay plan has been saved.

CASE STUDY

MARK LIU 'CLEVER PATTERNS CUT WASTE OUT OF FASHION'

Mark Liu is a graduate of the Textile Futures Course at Central Saint Martins College. He showed his first collection at the Estethica exhibition during London Fashion Week in 2008, where his ideas stood out for their beautiful fabrics and delicate detailing. These details however, were not an afterthought, but integral to the production of the garments he designs and makes.

Mark Liu believes that it is important to cut waste out of fashion and he does so by working carefully with his lay plan ideas, one pattern at a time. He certainly is good with a pair of scissors! Mark Liu's 'Zero Waste Fashion' mission was inspired by the

Figure 5.9 Mark Liu's designs are both beautiful and environmentally friendly

shock of seeing 15 per cent of all fabrics going to waste in the pattern cutting industry and he challenges himself with pushing tailoring to its limits. After all, wasted fabric leads to a loss of profit as well as being bad for the environment.

By designing and cutting pattern pieces like a jigsaw, from a single piece of fabric (as shown in Figure 5.9), Mark Liu creates designs without any waste. Mark Liu produces new forms and details in his clothing that are totally distinctive and, in today's fashion world, 'on the cutting edge'.

KEY POINT

- Using CAD and CAM means that pattern pieces can be cut accurately and efficiently in less time and with reduced fabric waste.

ACTIVITY

1. Using a commercial pattern of your choice as your starting point, sketch out a lay plan to show how the pattern pieces would be placed on the fabric to cut out a small batch.

2. Read the case study about Mark Liu. Choosing a simple pull-on hat pattern, cut out the pattern pieces and sew together using the same method as Mark Liu. Write down your thoughts about this process; does it reduce fabric waste and is the final effect successful?

5.4 METHODS OF LAYING OUT FABRIC AND CUTTING OUT

As a learner you will need to have knowledge about lay planning and the effective use of fabric to avoid waste. You will also need to have an awareness of how CAD and CAM are used to produce a pattern lay in industry.

Fabric should be folded according to the instructions on the pattern when you are working at school or at home. The right side of the fabric should be placed on the inside of the fold to help protect it from getting marked during preparing and cutting. The pattern pieces should be placed on the fabric as close together as possible, following the directions printed on the pattern. Care needs to be taken when working with a fabric which has a stripe, check, nap, pile, stretch or one-way direction, so that when the pieces are stitched together the design and pile match.

Notches can be used to line up the pattern pieces correctly to avoid this problem and pattern pieces need to be arranged to suit the design and characteristics of the fabric.

Computerised cutting

The marker or lay plan is put on the top of a single layer of fabric. A computerised numerically-controlled cutter follows the patterns outlined on the lay plan and cuts the fabric using a cutting head (mechanical, water pressure jet or laser). The movement of the cutting head is directed by the pattern lay data in the CAD system, which can also be viewed on the computer screen. This process is an integrated CAD and CAM system and is referred to as CIM (computer-integrated manufacturing).

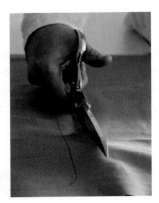

Figure 5.10 In industry toiles are usually cut out by hand

Advantages of industrial cutting tools are:

- will cut different thicknesses and types of fabrics accurately
- all pieces cut from the lay are identical in edge finish
- suitable for precision cutting – corners, tight curves and pointed incisions are cut precisely

- process is fast and economical
- quality is maintained over long runs
- a choice of cutting forms is available to suit any fabric type – vertical blades, laser blades, high-pressure water jets.

Disadvantages of industrial cutting tools are:

- expensive to install equipment initially
- some methods are slow and rely upon the skill of the cutter – straight knife cutter, band knife.

In industry, samples and toiles are usually hand-cut with shears, using the full length of the blade. This is the method you will use in the classroom to model your final item. Remember to pin the pattern securely to the fabric and hold the pattern pieces with the flat of your hand as you cut.

Preparation and marking of fabric

It is important that marks or notches used to match pieces together are not visible on the finished item. There are a variety of ways the marking process can be completed in industry:

- **drill markers** – make small holes in the fabric layers
- **dye markers** – where the holes are marked by a colour. This is usually fluorescent which can only be detected under an ultraviolet lamp. Pocket positions and darts are usually marked this way.
- **thread markers** – a tacking thread is stitched through the layers of fabric and is cut between each layer
- **hot notcher** – used for marking the edge of the fabric where notches may be seen. This method is used only on knitted or natural fabrics.

In the classroom you will usually use tailor's chalk or pencil, fabric marker pens or tailor's tacking to transfer markings on to your fabric pieces.

KEY POINT

- The use of computers has made the process of garment cutting and marking easier, faster and more accurate.

ACTIVITY

1. Using the resources you have available to you, research and present a report about each of the industrial marking methods listed in this section.

2. Explain how CAM can be used to cut out pattern pieces.

5.5 FINISHING

You will need to know about the different finishing processes used in industry. Finishing can involve a range of different tools and equipment to ensure that a quality outcome is achieved.

Pressing matters

In industry, pressing is carried out at all stages of the making process to ensure a high quality finish is achieved for every product. The two most effective methods used for pressing textile items are moulding and top pressing.

- **Moulding**: this method of pressing helps to enhance the appearance of the final item by concentrating on the areas which give shape to the product.

- **Top pressing**: this is the final pressing of the finished item where all the parts of the textile product are pressed.

Machinery has been developed to ensure that the pressing process is performed accurately and quickly.

Steam dolly

The steam dolly is used to steam-press the final garment at the end of manufacture. It is shaped like a body form which the completed garment is placed over. The dolly is inflated using steam and air, which removes creases and shapes the garment. The garment will then be checked and manually pressed in various areas to ensure a quality finish.

Tunnel finisher

The tunnel finisher is also used to press the finished item and is better suited to shirts and blouses. The garments are hung on a hanger and steamed. Manual 'spot' pressing of individual areas, for example the collar, is done with a steam iron to improve the quality of the final product.

Presses

The flat press and the moulding press are the most popular methods used in industry. These pieces of equipment press areas of the textile product as it passes through the production system. Steam, pressure, temperature and the length of time used determine the final finish. All of these can be adjusted to suit different fabric types and different textile items.

Figure 5.11 Manual 'spot' pressing in industry can improve the quality of the final garment

KEY POINT

● It is important to press a textile item at each stage of production in order to achieve a quality finish.

ACTIVITY

What industrial method of pressing would you use to finish a) a shirt and b) a dress? Give detailed reasons for your answer.

5.6 LABELLING AND TAGGING OF TEXTILE PRODUCTS

You will need to understand the labelling and tagging requirements for textile products, including the development of smart labelling and radio frequency identification (RFID) tagging for your written examination and aspects of your coursework.

Labelling

The labelling of a textile product for fibre content has been compulsory since 1986 with the introduction of the Textiles Products Regulations Act. This ensures that the user has all the necessary information to care for the product, and ensure that it remains fit for purpose for as long as possible.

Labelling of a textile product helps to ensure:

● information about fibre content is evident
● all the important information leading to the effective care of the product is available to the user

● the manufacturer conforms to European Union (EU) standards by keeping the names of the fibres and the method of labelling the same throughout the EU
● more effective stock control in retail outlets, through the use of a tagging system.

The labelling scheme is compulsory for all new (unused) textile products containing more than 80 per cent textile fibres. The retailer has the responsibility of ensuring that the products on sale are labelled: failure to do so may lead to prosecution.

The most common form of labelling a product is by a permanent label sewn along a seam, waistband or neck edge of a garment. Care labelling may also be found on a gummed label attached to the packaging or on a swing tag attached to the product.

Figure 5.12 Information on a permanent care label

Label information and symbols

Textile items are made and sold in many countries across the world. To ensure that every user understands how to care for the product a set of symbols, known as the International Care Labelling Code (ITCLC), has been developed. The care code is based around six basic symbols which are adapted to give the necessary information to the user.

EXAMINER'S TIPS

It is useful to remember that any symbol with an X through it means that you should *not* perform this process on the product. For professional cleaning and washing a line or double line underneath the symbol means clean gently or very gently.

Hand wash | Do not bleach | Do not tumble dry | Do not iron | Do not dry clean

Figure 5.13 The basic care symbols and their meanings

Care labels should provide the user with the following information:

- fibre content – if the product is made up of more than one fibre, the two most important fibres must be named with the percentage weight
- the size of the garment
- all care instructions – washing, drying and ironing
- points to note – whether the item needs to be dried flat or washed separately
- dry cleaning information
- manufacturer's details – country of origin.

It is important that the user reads and follows the advice given on the care label to ensure that the product retains its functional and aesthetic qualities for as long as possible.

Hot iron 200°; cotton, linen viscose

Warm iron 150°; polyester mixes and wool

Cool iron 100°; Acrylic, nylon, acetate, polyester

This symbol is used to indicate that ironing will damage the garment

Tumble dry at high heat

Tumble dry at low heat

Do not tumble dry

Figure 5.14 Ironing and tumble dryer symbols and their meanings

Smart labelling

Developments in technology have led to accurate and fast ways to label products in industry. The **Avery Dennison 676** is a high-resolution printer used to produce tickets, stickers and care labels on a range of nylon and polyester fabrics. Once printed the labels are cut to size and stacked, ready for sewing. Alternatively, for soft satin fabrics an **ultrasonic cutter**/stacker unit can be used. The advantage of using this cutting method is in the smooth, softer edge which is produced on the label, which reduces irritation to the user; the system is low-cost and is easy to use.

Figure 5.15 The Avery Dennison 676 high-resolution printer and ultrasonic cutter

▶ Tagging

The tracking and assessment of stock is an important process in industry to maintain a cost-effective system. Radio frequency identification (RFID) tagging is a system developed to enhance the existing bar code tagging process.

How does RFID work?

RFID tags are intelligent electronic 'chips' made for a specific textile item. They are programmed with unique identifier information such as the item description, item ID code, date of production and status information. This process is completed before the item is sent to the retail outlet. Once the tagged item leaves the manufacturer the process becomes entirely automated – an antenna device is used to read the tags as the garment arrives at the store and the system automatically updates the computerised stock control system. RFID tags can also be checked manually with a hand-held terminal.

Figure 5.16 Intelligent tag

Advantages of the RFID tagging system:

- the information stored on the tag can be changed or updated when necessary
- stock flow can be closely monitored
- the tag can be reused, which reduces production costs and is more environmentally friendly
- several textile items can be programmed or read together at any stage of the distribution process
- tags are designed to be flexible
- checks can be made on the movement of the product within the store until the tag is removed at the point of sale.

CASE STUDY

The outdoor clothing company Icebreaker, pioneered the Merino apparel market, taking tagging to the next level through 'Baacode', a revolutionary traceability system. The system allows the customer to follow their garment through every step of the production process: from farm through the entire supply chain.

Figure 5.17 Icebreaker's 'Baacode' technology makes garment tracking by the consumer a reality

KEY POINTS

- Fibre content labelling is compulsory by law in the EU.
- In industry labels can be applied quickly and efficiently using computerised systems.
- The tagging process helps to ensure a more effective stock control system.

ACTIVITY

Using CAD, create a care label for a) a swimsuit, b) a silk shirt and c) a woollen cardigan.

5.7 HEALTH AND SAFETY

In this section you will learn about:

- the responsibilities of designers and manufacturers to the workforce, the consumer and the general public
- the importance of personal safety when engaged in designing and making activities
- personal protective equipment
- machine guards
- dust and fume extraction
- waste disposal
- accident procedures
- risk assessment procedures
- COSHH
- safety standards symbols.

Health and safety are important aspects to consider when designing textile products. Legislation associated with the equipment, the working environment and the workforce are all key areas of importance. You will need to be able to understand health and safety as a designer, manufacturer and consumer.

KEY TERMS

DESIGNER: person who designs the product.
CONSUMER: the intended user of a particular product.
MANUFACTURER: producer of goods.

▶ Responsibilities

Designers and manufacturers have responsibilities not only to their workforce but also to the consumer and the general public.

Every employer is required to ensure that their employees work in a safe environment. Most companies employ a Health and Safety Officer to ensure that:

- All equipment, tools and machinery are safe to work with and have been tested for safety. Safety tests for equipment should take place yearly and equipment should be labelled and dated.

- All workers wear suitable protective clothing.
- A safe work environment is provided, e.g. correct temperature, the right lighting, not too noisy, no dangerous obstacles or trailing wires.
- All processes are safe and do not damage the health of employees.

Employees are also responsible for their own health and safety and those of fellow workers around them; all rules and regulations must be followed. This also applies to you in school and your fellow students. Health and safety labels, posters and warning notices should be prominently displayed in all working environments.

NO SMOKING

Fire Extinguisher

First aid

CAUTION Dangerous machinery

Figure 5.18 Warning signs

In your role as a designer or a consumer you need to:

- take responsibility for the correct selection of materials and finishes
- use safety information to help you assess risks involved in the designing, making and using of textile products
- ensure that any workers within the production environment are managed and working safely and effectively.

As a worker within the production environment you need to be aware of:

- **ventilation**: legal requirements so as to not be working with fumes, e.g. from dyes in a screen-printing factory
- **protective clothing**: e.g. when using dyes and chemicals, workers should wear overalls, goggles and Wellington boots
- **machine guards**: rules and regulations are set to reduce the risk of an accident, e.g. using a sewing machine in a factory – is the finger guard in place before sewing? If not a finger injury could occur.
- **accident procedure**: what happens if an accident occurs? Who is responsible?

Personal protective equipment

Protective clothing such as suitable goggles, masks, gloves, hard hat, footwear and overalls may need to be worn, depending on the task you have chosen to do.

Machine guards

Machine guards are used to prevent injury, mostly to fingers and hands. Most industrial textile machines have guards fitted to prevent hands and fingers being placed near moving parts. The guards often need to be removed to set up or maintain the machine, and it is important to check that they are correctly replaced afterwards. Following the correct instructions for the machinery in school is very important for the safety of the users.

Safety with people	Safety with materials	Safety with equipment
Tie hair back	Handle with care	Hands away from cutters
Wear protective clothing	Take care with hot liquids	Put tools away
Wear safety goggles	Wear protective gloves	Switch machines off
Follow safety rules	Clean up spills	Leave equipment clean
Sit correctly and at a correct height	Use correct tools	Check correct temperature e.g. iron
	Labelling of chemicals and dyes	Training to use the equipment
	Electrical checks	

Figure 5.19 Personal health and safety chart

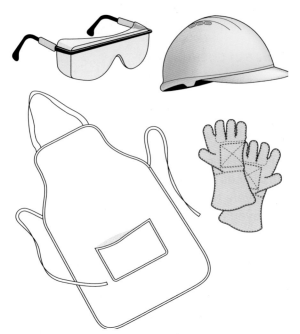

Figure 5.20 Protective equipment

Dust and fume extraction

There must be adequate ventilation to protect those working with processes that cause dust or fumes, such as screen printing with dyes or applying chemical finishes such as flameproofing. The particles in the air from fibres, dyes or chemicals can cause damage to the lungs and irritation to the eyes and skin. Extractor fans can be installed in areas used for these processes or separate areas can be designed such as 'fume cupboards' where these processes can be carried out. As with all extraction and ventilation systems, regular servicing and cleaning checks must be carried out to ensure the safety of the users.

Waste disposal

Manufacturers have to follow guidelines on how to dispose of their waste effectively.

This waste includes:

- dyes and chemicals used in processes
- water used in any process
- packaging.

Many manufacturers are involved in schemes to cut waste and are looking at more environmentally friendly systems. You can read more about this in Chapters 6 and 10.

Accident procedures

All workplaces, whether a factory or school, should display details of health and safety procedures. These instructions should include evacuation procedures in the event of fire. Often in the event of an accident or fire drill, paperwork must be completed to log what has happened, where it happened and the outcome of the event.

Risk assessment

The Health and Safety at Work Act 1974 was drawn up to protect employees from hazards at work. Everyone has a responsibility to ensure that the way in which they work causes no harm to others.

Risk assessment means identifying the risks to the health and safety of people and to the environment. It involves looking at all the activities carried out in the workplace and assessing the actual risks involved with each one. It is a legal requirement for all workplaces, including schools, to undertake a risk assessment. After a careful and detailed examination has taken place identifying an area, act, or item which could potentially cause harm, procedures and systems must then be put in place to eliminate, reduce and control the risks.

Risk assessment covers all aspects of the workplace:

- the layout of the workplace
- the environment of the workplace
- the use and storage of tools and equipment
- the use of machinery
- the use and storage of chemicals and dyes.

ACTIVITY

Create a list of equipment you have used to make a recent product. List the potential hazards or risks associated with this equipment and explain how these risks could be avoided.

Figure 5.21 COSHH symbol

COSHH

Control **O**f **S**ubstances **H**azardous to **H**ealth.

Using chemicals or other hazardous substances can put people's health at risk. Employers have a duty to protect both the employees and others who may be exposed. Exposure to hazardous materials can result in:

- skin complaints – rashes, itching and general soreness
- eye irritation
- fume inhalation – giving symptoms from headaches to fainting
- chronic lung disease
- fatality (although this is very rare).

For the employer the consequences are of a different nature and could include:

- loss of productivity

INDUSTRIAL PRODUCTION   137

- liability of legal action – including prosecution under the COSHH regulations
- worker disputes and action.

It is far more beneficial for an employer to work with the regulations and to ensure good working practice at all times. Effective control of hazardous substances can lead to improved productivity and better morale among employees.

Hazardous substances typically include adhesives, paints, dyes and chemicals. They also include certain cleaning materials and developing materials which can be both irritants and give off fumes. It is important to remember that COSHH only applies to products if they have a warning label, for example fabric dyes.

Toxic

Irritant

Figure 5.22 Warning symbols

▌ Safety standards symbols

Some products are required by law to include health and safety warnings. Products must be made from safe materials that are non-toxic for people and the environment and must be safe to use. This information can be shown in a variety of symbols that have become easily recognisable. You will need to be able to identify and name these symbols. Figures 5.22–5.26 show some important symbols.

Figure 5.23 CE label

CE

The CE marking shows that the product meets the relevant EU (European Union) directive for safety relating to the product. It is used on products sold across Europe to signal that they have met the European legal, technical and safety standards. If it is used on a product that does not meet these requirements, the manufacturer can be prosecuted.

Figure 5.24 BSI kitemark

BSI kitemark

The BSI Kitemark is the world's best-known product and service certification mark. BSI stands for British Standards Institute.

Figure 5.25 The BEAB mark

BEAB

In the workplace, the BEAB (British Electrical Approvals Board) mark of safety is used to show that electrical goods have been tested annually to ensure safe working.

Sustainable labels and environmentally friendly product labelling are an important part of health and safety checks and will be looked at in detail in Chapter 6.

Hazard symbols

Hazard symbols can be found on various items used in the textile industry and can carry various warnings. Figure 5.26 shows the hazard symbols.

A range of hazard signs is used in the textile industry and workplace. Some signs are installed next to or on machinery, some can be found on packets or chemical containers.

ACTIVITY

Look around at school and see how many different signs or symbols you can find. Make a list of where you find them and what hazard they are identifying or warning of.

Biological hazard

Safety electrical haxard

Explosives

Harmful/irritant

Flammable

Corrosive

Figure 5.26 Hazard symbols

CHAPTER 6

ENVIRONMENTAL AND SUSTAINABILITY ISSUES

By the end of this section you should have developed a knowledge and understanding of:

- The 6Rs
- Globalisation
- The impact of the production of fabrics and fabric products on the environment
- Selection of materials: the use and disposal of fabrics and fabric products
- The life cycle of textile products

Environmental and sustainability issues are of key importance when both making your textile product and writing about it. Additionally, you will be examined on these issues in both of the textile examination papers.

6.1 THE 6Rs

Recycle

KEY TERM

RECYCLE means to reuse a product.

Recycling is the conversion of waste products into new materials, a creative act that involves thought to extend the life and usefulness of a product, item or object that seems to have no more purpose or use, once it has been finished with or used for its initial purpose. Recycling means reusing a product, but sometimes before a product can be reused it will need to undergo processing or treatment.

The three main types of recycling are:

Primary recycling

This is the second-hand use of textile products and refers to clothing, household textiles or decorative textiles that can simply be used again. Charity shops stock a large selection of recycled textile products. Giving items to friends and relatives or selling them on internet market sites are ways of primary recycling.

ACTIVITY

You can research secondary recycling and list the number of different materials/fibres that you can find that have been used in an imaginative way. A good starting point is denim fabric.

Figure 6.1 **Recycled clothes**

Secondary or physical recycling

This is the process in which waste materials are recycled into different types of textile products. The change the product will go through depends on its main fibre or material. Some products can be left to biodegrade before being regenerated into something else. A woollen jumper can be unwound and reknitted into another garment, for example a scarf.

Figure 6.3 **Stack of denim jeans**

Tertiary or chemical recycling

Products are broken down and reformulated; for example, plastic bottles can be recycled into fibres and then respun into polyester to make fleece fabric used for coats and blankets.

▶ Why recycle?

Everything we dispose of goes somewhere, although once the container or bag of rubbish is out of our hands and out of our houses we forget it instantly. Our consumer lifestyle is rapidly filling up rubbish dumps across the world and, as this happens, concern for the environment grows. When designing and making a new product, designers and manufacturers need to consider how their product can be recycled at the end of its life cycle.

Figure 6.2 **Woollen scarf**

You will need to know the following:

- materials and fibres that can be recycled
- textile products that use recycled materials
- disassembly – reprocessing materials for use in new products.

Recycling through second-hand clothing has always been popular in the UK. Today many vintage shops have opened alongside the existing charity shops. Unwanted clothes are also exported from the UK to less developed countries. This is a growing trade, with other countries purchasing bales of clothes to distribute to shops throughout their country.

Recycling old clothes and remaking them into other garments or craft items is also popular. Patchwork is a particularly popular technique when sewing with recycled textiles. Items made from patchwork include bed throws, jackets, bags and hats.

Reuse

Reuse for either the same purpose or a new purpose
Items that are designed to be reused result in less waste, which leads to conservation of materials and resources. Many places around the UK collect unwanted textile products (clothes banks) or repair them for redistribution for the same or a similar end use.

Reuse – products that can be adapted for an alternative use
An example of this would be the alteration of clothes to make other items of clothing or household items such as cushions.

Some areas have set up their own websites and organisations for the reuse of unwanted items. These organisations aim to adapt existing products for alternative uses.

Figure 6.4 Student worksheet showing reuse of denim products

Reduce

A new product progresses through a variety of stages from conception of the idea to its decline, where it might be discontinued or disposed of. You must consider the impact of a textile product on the environment and its impact on society as a whole.

Think about ways you can reduce your waste. Changing your own habits is the key. What you buy, what you use and what you throw away can all be reduced in some way. Creating new fashion accessories out of unwanted items can help the life cycle of a product.

- **Built-in obsolescence**
  This refers to a product that has been designed to last a set period of time. The functions of the product have been

Figure 6.5 Shoes made from bus seat covers

designed by the manufacturer to fail after a certain time limit. The consumer is then under pressure to purchase again.

- **Energy and materials waste due to production processes**
  You need to be aware of the amount of waste created both in the process of making a product and in the actual materials used to make the product. Minimising the amount of material and energy used during the whole of a product's life cycle can make a big difference. The fashion industry is guilty of wasting energy through its choice of methods used to design and make garments.

▶ Refuse

Processing, manufacturing, packaging and transport of our products use huge amounts of energy and can create lots of waste. You need to look at the sustainability of a product from environmental and social viewpoints. How is the product made and can we ensure that little or no harm is done to the environment by this method of manufacture? Sometimes a choice between the performance required of the product and the impact on the environment by its manufacture has to be considered and debated.

Clothing prices today are now so low that many shoppers buy to wear and then throw away items of clothing after just a few occasions. Due to the low production costs in Third World countries, huge savings can be made by manufacturers during production. These cost reductions can be passed on to us – the consumers. Workers in these countries often work for low pay and in poor conditions. What do you think about this? Should we refuse to buy such products when we know how they have been produced?

Figure 6.6 Reduce waste

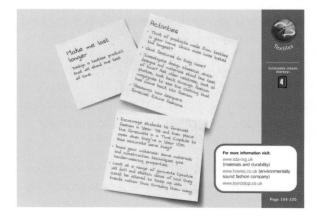

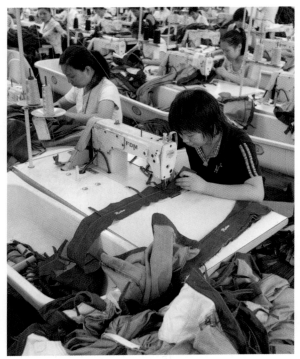

Figure 6.7 Factory sweatshop in Mexico

Rethink

Within your own lifestyle you need to rethink the way in which you buy products and consider the energy required to create them. We have become a throwaway society. We need to think about the huge waste sites that are blots on the landscape, both ugly to look at and unhygienic. These sites create other problems as they grow. As society evolves and our needs change, what can you do to rethink your part in what is happening both locally and globally? How could you make a difference?

Repair

This is about the throwaway society of today, where it is quicker and easier to throw something away rather than repair it. You have looked at built-in obsolescence earlier in this chapter, whereby manufacturers encourage consumers to repurchase rather than repair. We could all repair a product, whether it is an item of equipment or a textile product. Can repaired clothes be cool? Many years ago, most of the clothes worn by people in everyday life would have been repaired many times over, from shirt collars to patched jumpers and darned socks.

ACTIVITY

Write a list of the clothes you or family members have bought in the last two years.

- Have you still got them?

- Can you remember exactly what you have bought?

- What items do you no longer have? Why not?

How well have you done? Have you repaired any of your clothes or did you just discard them? Questions like this can make you think and be more aware of what is going on around you. To take this further you could try asking people of different ages to see if that makes a difference to your findings.

6.2 GLOBALISATION

Globalisation is the term used to describe the manufacture of textile products across the world. Technology and communications systems have improved substantially, enabling economic markets around the world to compete for business. The effect of globalisation on the textiles industry can be clearly seen. An example of this would be a sportswear manufacturer which designs its products in Europe has them made in South East Asia and then sells them in North America.

Understanding the impact of this global production and the use and disposal of fabrics and fabric products on the environment is very important. Social and moral implications must also be considered.

The Fairtrade Foundation is an organisation that seeks to ensure greater equity in international trade. The FAIRTRADE Mark attached to textile products shows that the product has been produced in accordance with internationally agreed Fairtrade standards. The workers receive a fair wage for their products and working conditions are supervised.

CASE STUDY
SO INDIGO

so **indigo**

So Indigo is a Cornish-based small company of two women who became interested in the idea of promoting and selling sustainable fair-trade-produced textiles. They import a range of textile products such as children's toys, puppets, bags, purses, floor mats and beach mats. The products are bold and exciting and are never mass produced. They are imported from a company called Barefoot in Colombo, Sri Lanka.

All the products are made from 100 per cent woven cotton, cotton and silk or wool yarn and each is handmade by one person. The dyes are eco-friendly and safe and where there is filling used, it is kapok pulp. The fibres are natural and therefore biodegradable.

Ethical trading standards ensure that the people making the products are given training, a decent wage and good working conditions. Products are produced in either small rural workshops or the home. All the toys featured on the So Indigo

Figure 6.8 So Indigo

website carry the European safety symbol or a CE label.

Products when posted and packaged use recycled stationery and packaging, thus ensuring that if non-biodegradable packaging is sent to the company they can ensure that at least it is recycled. Where possible, all postal bags are biodegradable. Imported goods are sent by sea to lessen the carbon footprint.

For more information on this individual company the website is: www.soindigo.co.uk.

Many famous designers today are also actively involved in the promotion, design and making of socially responsible textiles. These are textile products that use eco-friendly resources, that are sustainable and produced with socially acceptable and moral standards.

Katherine Hamnett is one such designer who works with a 'Go Green' ethos, designing clothing that is both fashionable and wearable and made from fabrics that are organic, recyclable and coloured with dyes from natural sources such as vegetables.

CASE STUDY
SARA SIMMONDS

Designer Sara Simmonds produces a range of luxury jeans with an ethical label, 'Sharkah Chakra'. Her handmade jeans are made in India, using organically produced cotton which is then dyed in natural indigo and left to dry and bleach in the sun. The cotton is then handwoven at the homes of village craftsmen. The entire process takes six months compared to six weeks for mass-

produced denim. One major benefit of the business has been the return of traditional crafts, which have turned around the lives of entire villages and families. It has enabled ancient skills of dyeing and handweaving to be taught to next-generation workers. Each pair of jeans is unique and carries a nine-carat gold part-recycled rivet to hallmark its authenticity. They also come with a friendship

Figure 6.9 Sharkah Chakra craftsmen

Figure 6.10 Handmade jeans

bracelet made by the workers. The retail price is more expensive than mass-produced denim jeans at about £195.00 per pair. These are not as expensive as some designer brands. For more information on this company look at their website: www.sharkahchakra.com.

6.3 RESOURCES AND THE PRODUCT LIFE CYCLE

Resources

It is important that designers protect natural resources. Fibres and fabrics can come from managed resources such as forests or crops. A managed resource ensures that as a plant or tree is harvested a replacement is planted.

Manufacturers can also research and use alternative energy sources, such as wind farms for producing electricity. They also need to be aware of transportation and packaging costs of their products. A carbon footprint is the measure of a manufacturer's energy uses in terms of manufacturing processes and transportation of both raw materials and finished product.

Managing waste is another issue that manufacturers have to address. Government guidelines are used to inform companies of what is allowable and what is not. Waste dye and waste water are the key areas of concern in textile manufacturing. This can be a particular problem in Third World countries where environment laws are less strict, non-existent or not enforced and there are often news reports about waste dye products being drained into natural water sources and polluting them. Some textile companies have their own guidelines in place as part of what is called their 'social responsibility document'.

Product life cycle

What do we mean by a product life cycle?

Product life cycle is a term used to describe the stages that a new product goes through from the initial concept to eventual disposal or decomposition of the product. A product life cycle assessment (LCA) means assessing the effect a textiles product has on the people or the environment. It means investigating every aspect of the product design, manufacture, use and disposal. Questions that you might ask to determine the impact your textile product might have are:

- Are the raw materials renewable?
- How much energy has been used during manufacture?
- Does the manufacturing process cause risk to people or the environment?
- Is the textiles product safe to use?
- Can the product be disposed of safely?
- Does its disposal cause risk to people or the environment?
- Can the product be recycled?

When we buy a product it is important to try and consider environmental and sustainability issues. The more effort each and every one of us makes, the better the prospects for the environment. Whether it is the fabric you are selecting to use or the use of a recycled component, it is the little details that can also make a difference. Try and refer to the 6Rs where possible to help you create and design sustainable and eco-friendly products.

ACTIVITY

Areas you could develop in your design-and-make units:

- Produce a design brief that has a recycled element to it.

- Make your own design folder from reused or recycled materials.

- Keep a record of your use of energy. Produce a time chart of time spent using electrical equipment.

CHAPTER 7

QUALITY

**By the end of this section you should have developed
a knowledge and understanding of:**

- How to distinguish between quality of design and quality of manufacture
- How the quality of a product may be affected by materials and processes used in its manufacture
- The importance of accuracy when making textile products
- How to generate designs for templates and patterns to control accuracy in batch production of textile products
- A range of simple quality-control checks to ensure accuracy and quality of finish

*You will need to have a clear understanding of the meaning of the term 'quality'
and how the management of quality impacts upon the successful production of
a textile product which is fit for use by the consumer. There are two key areas
to consider when assessing a textile item for quality:*

- *how well the product has been designed to fit its purpose (quality of design)*
- *how well the product has been made to meet that purpose (quality of manufacture).*

*It is important to remember that a quality textile product is one which has been
well designed and well made to meet its intended purpose.*

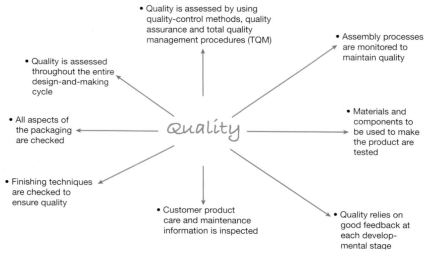

Figure 7.1 Ensuring quality throughout production

7.1 QUALITY OF DESIGN AND QUALITY OF MANUFACTURE

Quality of design

This is dependant upon the market for which the product is intended. For example, textile items like a bag, which can be made in a range of different qualities to suit a range of consumer needs and price brackets.

Quality of design is linked to how attractive a product is to its target market, how well chosen its materials and components are and how easy the product is to make and look after. Quality-control checks are usually made at the following points of the design stage:

- The product's **design specification**, which includes details of a product's required characteristics and all the processes, materials and other information needed to design the product. How does the product measure up to its purpose?

- The product's **aesthetic qualities** in relation to shape, style, colour, pattern and other aspects of visual appeal.

Quality of manufacture

Quality of manufacture is easier to define and looks at how well made a textile product is. A high-quality textile product will have the following characteristics:

- materials used are of good quality and suitable for the end use

- the product will match both the design and product specifications

- it will meet performance requirements, i.e. will be durable, waterproof, etc.

- manufacturing of the product will be completed using safe production methods

- the product is made within a budget limit

- codes of practice have been followed during manufacture

- the product is environmentally friendly

- the product conforms to relevant safety or moral issues.

Critical control points (CCPs) are areas in a manufacture system where quality-control checks take place. Figure 7.3 outlines some typical CCPs in textiles manufacture.

Before the final manufacture of a textile product, **a prototype** or mock-up of the product is made to trial a design or pattern and to see how materials and components perform. The making of the prototype also allows the manufacturer to try out an assembly process, work out costs, and test the products fitness-for-purpose in everyday use.

KEY POINTS

- Quality of design and quality of manufacture are different aspects of the production process, but they are dependant on one another in the making of a quality textile item.

- Quality of design is linked to how attractive a product is to its target market, whereas quality of manufacture means how well made a textile product is.

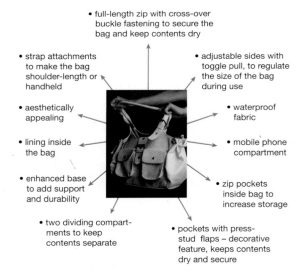

- full-length zip with cross-over buckle fastening to secure the bag and keep contents dry
- strap attachments to make the bag shoulder-length or handheld
- adjustable sides with toggle pull, to regulate the size of the bag during use
- aesthetically appealing
- waterproof fabric
- lining inside the bag
- mobile phone compartment
- enhanced base to add support and durability
- zip pockets inside bag to increase storage
- two dividing compart-ments to keep contents separate
- pockets with press-stud flaps – decorative feature, keeps contents dry and secure

Figure 7.2 The features on a bag must be *designed* to suit the needs of the user

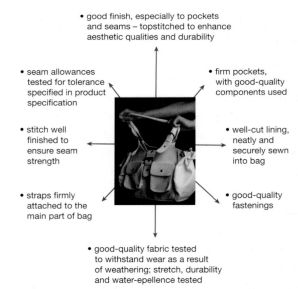

- good finish, especially to pockets and seams – topstitched to enhance aesthetic qualities and durability
- seam allowances tested for tolerance specified in product specification
- firm pockets, with good-quality components used
- stitch well finished to ensure seam strength
- well-cut lining, neatly and securely sewn into bag
- straps firmly attached to the main part of bag
- good-quality fastenings
- good-quality fabric tested to withstand wear as a result of weathering; stretch, durability and water-epellence tested

Figure 7.3 The features on a bag must be *manufactured* to meet the needs of the user

7.2 QUALITY ASSURANCE

Quality assurance is making certain that a textile product meets the quality standards set and guarantees this quality to the consumer. Legislation has been introduced to ensure that the manufacturer keeps this 'promise'.

The British Standard 5750 UK National Standards for Quality Systems is a scheme introduced to ensure that manufacturers 'get things right the first time', and which prompted the initiation of total quality management (TQM). TQM is covered in detail later in this chapter.

International Standard ISO 9000 is a testing system which ensures that the manufacturer produces products that are fit for purpose and the intended market. The BSI kitemark symbol is used to assure the consumer that rigorous tests have been applied to the product and to check safety in production and use. This applies only when the manufacturer's instructions are correctly followed by the consumer.

European 'CE' mark is the European safety standard symbol which needs to be shown on all clothing products that claim to 'protect' the wearer; this applies to work and leisure clothing. The European CE mark also indicates that a product, e.g. a toy, has passed mechanical, physical, flammability and electrical tests.

Total quality management (TQM)

Textile companies need to be able to guarantee that each textile item they produce is of the quality stated in the product

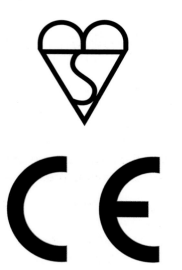

Figure 7.4 The BSI kitemark and the European safety standard symbol are a visual guarantee to the consumer that quality procedures have been applied

specification. To ensure this, the company has a set of quality management procedures that constantly seek to improve the performance and quality of each manufacturing process. This is known as **total quality management**, where each member of the company understands the importance of what they are doing and that they do it to the highest possible standard. By applying these procedures the manufacturer can ensure that:

- faults are identified at an early stage and rectified to ensure that faulty goods are never sold

- records and documentation are kept up to date and available for inspection

- a high-quality and detailed specification is produced, to ensure all materials and processes are of a consistent quality

- the consumer has confidence in the products produced, which ensures repeat orders.

7.3 QUALITY CONTROL

Quality control is a set of checks to ensure that a product is being made to the correct specification. Assessments are made to test and monitor the quality, accuracy and fitness for purpose of the product, from the design stage through to manufacture.

These tests give vital feedback information about a product, materials and components and the processes used. This is useful in giving guidance to the manufacturer if the product fails.

The quality of the final product depends upon a variety of factors, including the quality of the raw materials, the quality of the assembly process, the quality of control checks and the quality of the finishing techniques used.

Fabric quality

How can you ensure that you buy or make a quality textile product from quality raw materials? Follow these helpful hints:

- Always check the fabric content to see how to care for it to keep it looking good.
- Let your sense of touch be your judge – fabric needs to feel pleasant inside and out.
- Squeeze a handful of fabric for crease reaction.

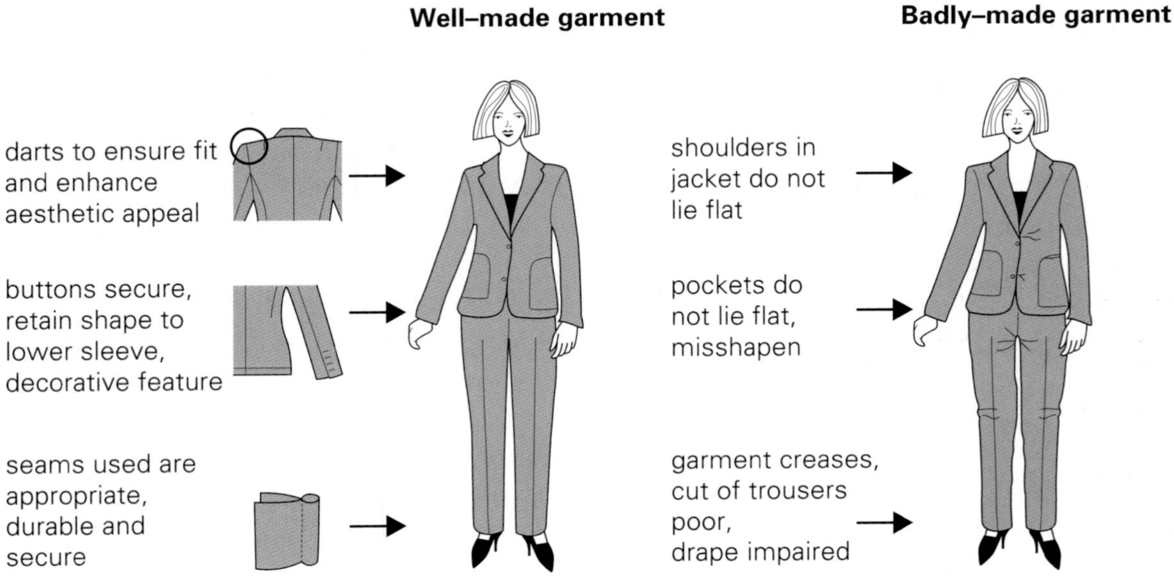

Well–made garment   **Badly–made garment**

darts to ensure fit and enhance aesthetic appeal

buttons secure, retain shape to lower sleeve, decorative feature

seams used are appropriate, durable and secure

shoulders in jacket do not lie flat

pockets do not lie flat, misshapen

garment creases, cut of trousers poor, drape impaired

Figure 7.5 The quality of a product may be affected by materials and processes used in its production

- Before buying fabric, ask yourself if it is really something you need and like.
- Fabric that drapes smoothly flows smoothly.
- Watch out for fabrics that may shrink, expand or discolour. (Cotton, linen, rayon may shrink or expand. Silk, cotton, linen may discolour or bleed.)

Quality-control checks

To make sure that a textile product meets the specification requirements, it is put through a series of rigorous tests at different stages of production. The tests may include:

- checking the accuracy of the dimensions and all tolerances
- checking of yarn to ensure uniform thickness and other characteristics
- fabric testing according to purpose, e.g. flammability, durability, stretch and recovery, washability
- looking at the fabric to ensure there are no faults in its construction, e.g. slubs or marks
- ensuring that all manufactured components, i.e. buttons, etc., are fault-free, of the same size and fit for use
- checking manufacturing processes, e.g. seam construction and tolerances
- visual checks to ensure that appearance and colour match the specification
- checking that the product conforms to the correct standards legislation.

Quality control in industry

In industry, accuracy and quality are usually monitored by a **clothing/textile technologist**

KEY POINT

- Rigorous quality-control checks during the design and manufacturing stages of a textile product ensure accuracy and quality of finish.

who carries out a range of technical, investigative and quality-control work on clothing and textiles, ensuring that products perform to specifications. ICT experts can also be utilised by the manufacturer, using CAD and CAM to automatically check production processes to ensure an accurate, quality outcome.

Figure 7.6 The quality control department runs some checks on the fabric

In what ways can CAD/CAM be used to control accuracy in batch production?

There are many ways that CAD/CAM is useful to the manufacturer to increase accuracy and efficiency.

- Garment patterns are developed from a basic block using CAD (computer-aided design), a system which helps the manufacturer produce accurate drawings (see Chapter 5).

- Texture mapping 3-D computer-modelled prototype garments by using a computer-aided design and automation system, allows for more accurate and consistent texture mapping of surface details on to a garment (see Chapter 4).

- Once a design has been approved it is put into production in a range of standard sizes, which can be controlled using templates. Templates ensure that the same things are placed in the same position on each item, for example, a buttonhole.

- CAM (computer-aided manufacturing) involves the use of CNC (computer numerical control), which uses computers to control cutting, printing and shaping machines and many other textile processes. CNC-automated machines can repeat processes with accuracy and reliability, and are easily reprogrammed when changes to the design or the production run are needed.

ACTIVITY

1. Product quality depends upon several factors. List *five* and give detailed reasons for your choice.

2. Describe *four* quality-control checks used in industry to ensure accuracy and quality.

3. Explain the difference between quality control and quality assurance.

4. Choose a picture of a textile product. List all the control checks needed to complete this product. Present your findings in a report created using ICT.

KEY POINT

- CAD/CAM or other manufacturing aids can be used to make a batch of identical parts or items easily and accurately.

CHAPTER 8

PRODUCT EVALUATION

By the end of this section you should have developed a knowledge and understanding of:

LEARNING OUTCOMES

- The function of commercially manufactured textile products
- How to evaluate specific materials for specific tasks
- How to evaluate textile products designed to meet the same need
- How to critically evaluate the making process
- Successful testing to determine fitness for purpose
- How to identify improvements to the design, materials and processes

To evaluate means to 'examine and judge an outcome carefully'. Being able to examine processes, systems, objects and ideas in terms of their component parts, and make informed judgements about their worth is an important part of the design-and-make process.

Evaluating your textile outcome is an integral part of both of the controlled assessment units. It is not just the final product that needs to be evaluated; the making process also has to be considered. This 'reviewing and refining' of your ideas and processes aims to improve the quality and efficiency of the systems you have used.

8.1 EVALUATING THE FINAL OUTCOME

The best way to start the review process is to evaluate your outcome against your specification criteria, using each of these points to identify how well each requirement has been met.

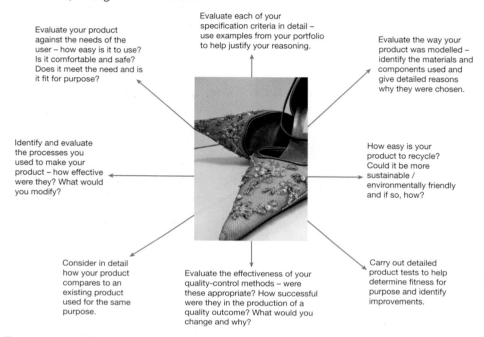

Evaluate your product against the needs of the user – how easy is it to use? Is it comfortable and safe? Does it meet the need and is it fit for purpose?

Evaluate each of your specification criteria in detail – use examples from your portfolio to help justify your reasoning.

Evaluate the way your product was modelled – identify the materials and components used and give detailed reasons why they were chosen.

Identify and evaluate the processes you used to make your product – how effective were they? What would you modify?

How easy is your product to recycle? Could it be more sustainable / environmentally friendly and if so, how?

Consider in detail how your product compares to an existing product used for the same purpose.

Evaluate the effectiveness of your quality-control methods – were these appropriate? How successful were they in the production of a quality outcome? What would you change and why?

Carry out detailed product tests to help determine fitness for purpose and identify improvements.

Figure 8.1 Evaluating the final outcome

8.2 WAYS TO EVALUATE YOUR PRODUCT

When you are evaluating your final piece of work, which consists of a portfolio and a final textile product, you should consider how you can present this information. The evaluation does not need to be a written section, but should include notes, sketches, photographs, opinions from users, diagrams, modelling and, where appropriate, graphs and charts.

What is meant by the term critical evaluation?

For your controlled assessment units A571 and A573 you will need to produce a critical evaluation where you will be expected to:

- critically evaluate the processes involved in making a final product/prototype
- reflect and suggest modifications to improve the making process.

Critical evaluation is about proving a point, interpreting information and resolving problems. It is the ability to make informed judgments or evaluations about the worth, validity and reliability of ideas, processes and knowledge. You will be expected to:

- test your results, taking into consideration the data collected and the processes you have used

- predict/evaluate alternatives to materials, components and processes
- collect, analyse and organise information
- evaluate and monitor your own performance
- explore alternative ideas or solutions to problems
- identify opportunities not always obvious to others/user.

Testing and trialling

The specification document asks you to 'test your own and commercially manufactured textile products to determine fitness for purpose and identify improvements to the design and materials and processes used with reference to innovation, environmental and sustainability issues'.

In industry, textile products are tested and evaluated at regular stages in the production process. This ensures that materials, components and production methods used are performing to the specification criteria. This is known as continuous evaluation.

In the classroom, you do not always have access to industrial evaluation and checking equipment or opportunities. Therefore, when testing and trialling your ideas, materials and the final product, you will rely heavily upon what you see and feel and how your user group reacts to your final outcome.

Suitable trialling methods could include:

- taking photographs of your product in use, questionnaires and surveys completed by the user which can then be analysed
- the use of a product analysis, specifically used to compare your final outcome with an existing product used for the same function.

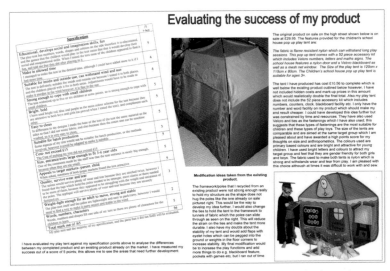

Figure 8.2 How a student has compared their own end product to one designed to meet the same need

8.3 EVALUATING MAKING

Evaluating the processes involved in making your final product or prototype (system evaluation), means that you look at the ways you could improve this process from pattern lays through to product completion. In industry system evaluation is carried out in the following ways:

- All the stages of the making process are listed using a flow diagram/chart. You must also remember to photograph each stage of making for your portfolio.
- Problem areas are highlighted and compensated for in the production run.
- Assessments and future improvements to the production process are suggested.

- Alternative ways of making/producing the product are considered – use a different system, improvements made to health and safety, implications for society considered, etc.

This process is useful to help you to produce a successful product evaluation in a classroom situation. You could also try remodelling parts of the final item in your evaluation to illustrate improvements to construction methods, aesthetic appeal, function, etc. Digitised imagery showing changes and modifications to specific areas could also be included.

EXAMINER'S TIPS

Remember to:

- **Evaluate** – produce a detailed evaluation of your product and the processes involved in making the final prototype. Evaluate your strengths and weaknesses and illustrate how you have considered the needs of the user.
- **Reflect** – consider carefully how the materials, components and processes you have used have 'combined' to form a quality product which is fit for purpose. Be sure to refer to the working properties of the materials to support your reasons.
- **Modify** – outline realistic and detailed modifications to your work. Use annotated sketches and modelling to support your points. Improvements should be creative and refer to innovation, environmental and sustainability issues.

ACTIVITY

Take a photograph of a textile product you have designed and made and an existing product used for the same purpose. Compare the two products, commenting upon the following:

* appearance/aesthetic qualities

* anthropometric qualities – size, ease of use

* ease of care – materials, components used

* cost

* environmental and sustainability issues – can the products be recycled?

* innovation.

Present your results in an interesting and informative way. You must make sure that your report includes a digital photograph of the two products.

CHAPTER 9

UNIT A571: INTRODUCTION TO DESIGN AND MAKING

LEARNING OUTCOMES

In this section you will learn about:

- The assessment requirements of this unit
- The evidence to be presented within the portfolio
- Options for presenting the portfolio

This controlled assessment unit aims to introduce the learner to designing and making through textiles technology. You will be able to select one of the themes published by OCR to use as a starting point for this unit. Once you have selected a theme you will need to identify a specific product or starting point associated with this theme. For example, if the chosen theme is 'eco-wear' you may decide to design and model a textile product made from recycled fabric and components from an existing product, or identify an existing product, such as an old pair of jeans, as your starting point to create your piece of eco-wear.

EXAMINER'S TIPS

It is important to remember that the themes will be reviewed every two years and that you need to select a different theme for Unit A573. Your teacher will be able to provide you with a list of the current themes.

9.1 STRUCTURE

Unit A571 makes up 30 per cent of the total GCSE marks (60 per cent for the short course). It is a 20-hour controlled assessment portfolio with a total of 60 marks available. The work associated with this unit must be completed under your teacher's supervision. However, some of your research work and testing of the product, by their very nature, may take place outside school under limited supervision. You must reference the source of any information you use within your portfolio.

Submission of controlled assessment work

Your work can be submitted on paper or in electronic format but not a mixture of both.

Work submitted on paper

A contents page with a numbering system should be included to aid organisation. All your work should be on the same size paper but it does not matter what size paper you use. You can produce your work by hand or by using ICT.

Work submitted electronically

Your work will need to be organised in folders so that the evidence can be accessed easily by a teacher or moderator. This structure is commonly known as a folder tree. There should be a top-level folder detailing your centre number, candidate number, surname and forename, together with the unit code A571. The next folder down should be called 'Home Page', which will be an index of all your work. The evidence for each section of the work should then be contained in a separate folder, containing appropriately named files. These files can be from the Microsoft® Office suite, movie, audio, graphics, animation, structured mark-up or text formats. PowerPoint® is ideal for producing electronic portfolios.

9.2 UNIT REQUIREMENTS

This unit will require you to produce the following:

- A number of concise worksheets (A3, A4 or digital equivalent) showing design development and modelling, which may include the use of ICT.

- A prototype final product, capable of evaluation. (A prototype is defined as the first example of a product that could be further developed or modified.)

- A minimum of *two* digital images/photographs of the final product, showing front and back views.

- Digital images/photographs of any models or mock-ups used when designing, modelling and testing.

- A completed OCR cover sheet.

You will be expected to include *only* the following areas within your controlled assessment portfolio:

- research and investigation skills
- drawing skills
- use of modelling

- evidence of production of a prototype textile piece
- an evaluation of the making process.

9.3 GUIDANCE POINTS

In order to skilfully design, model, make and test your prototype and complete your supporting portfolio, you need to read the following guidance points based upon an OCR theme.

Theme: *Eco-wear*

Starting point/outline:
'Recycling of textile items is an ever-growing trend within society. Create a new innovative piece of eco-wear by reusing the fabric and/or parts of an existing product.'

▶ Cultural understanding (8 marks)

In this section you will need to identify how designing and making reflects and influences culture and society, and look at how existing products have improved lifestyle and choice.

Write down the theme and your starting point clearly and use these to identify 'areas of action' through a mind map or thought diagram. This will help you to think about the key stages needed to complete your portfolio successfully.

Research

Research a range of existing products that have been designed around the theme of recycling. Consider how and why these

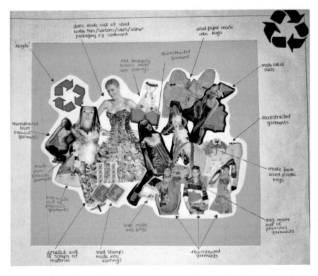

Figure 9.1 A student example showing initial thoughts related to the starting point

products have been produced how they reflect the changing trends, different cultures and different user groups within society. Link this to growing environmental awareness and sustainability issues. Use the information in Chapter 1 of this textbook to help you.

EXAMINER'S TIPS

- Remember to use appropriate textile examples, keep a note of websites and resources you have used (research and quotes should be 'acknowledged for source' next to each piece when written, *not* presented as a bibliography at the end of the portfolio) and present your information with flair and thought.
- Mood boards can be used but make sure they are justified and reasoned.
- Key words for this section – 'reflect, influence and culture'.

To identify and compare how textile products can improve lifestyle and choice, you could complete a **questionnaire** designed to analyse improvements made to lifestyle through fabric properties (increased comfort, easy care) and product design (aesthetic appeal, better fit). You will also need to consider the choice factor (cost benefits, range of sizes, colours, styles) in relation to user groups.

Design brief

At the end of this section you should be able to show that you have considered and researched the starting point in order to write a design brief. An example of a brief written in response to the theme 'Eco-wear' could be:

> *'I shall design and create a new eco-garment suitable for a teenager, by reusing the fabric and/or parts of existing textile products.'*

EXAMINER'S TIP

- Remember to use the key words – 'lifestyle' and 'choice'.

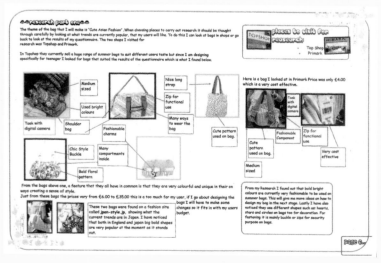

Figure 9.2 Student example of a comparison of textile products linked to lifestyle and choice

Creativity (8 marks)

In order to show that you understand how and why products have been developed to satisfy your user needs, a **product analysis** is required. You should compare three or four products. The products you analyse should satisfy a similar need and be based around the theme. It could be useful to explain your ideas in this section.

You will also need to identify existing products (from home or charity shops) which are available to you to recycle and remodel into a new and innovative textile piece. Show and explain your selected products, outlining where they are from, what details interest you and how the fabrics and components can be reused. You should link your reasons to fabric/component properties and performance.

Present your findings in an imaginative way: use digital technology, annotated diagrams, modelling, sketches, etc. to demonstrate specific details or areas of interest within a product which you would like to develop further.

You will need to collate and analyse your findings from the research you have carried out in relation to how culture, choice, lifestyle and society impact on the design of an existing recycled textile product. This analysis should be based around the principles of good design. For example, conclude your results from your questionnaire, product analysis and research by discussing the advantages and disadvantages of your findings.

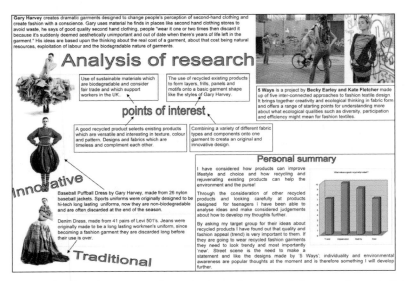

Figure 9.3 Student example showing research analysis and appropriate use of specialist terms

Designing (13 marks)

In this section you need to 'demonstrate an appropriate and considered response to a brief and produce a detailed specification for a textile product'.

This means that you need to show that you have considered and researched the starting point and the brief by writing a paragraph explaining your final findings and what you hope to develop further once a specification has been formulated. You then need to write a detailed specification. Chapter 1 will give you more guidance on how to write your specification.

EXAMINER'S TIP

- Make sure that you justify each specification point and consider environmental and sustainability issues.

EXAMINER'S TIPS

- Some questions you could ask are: What data have you found to illustrate what makes a good recycled product? Which points of interest would you like to develop further in your designs, and why? What type of product will you be looking to develop and investigate further?
- You will be able to gain extra marks for this section by showing that you have a good knowledge and understanding of textile terms and you have considered how to present the information gathered in a structured and imaginative way.
- You also need to show that you can use spelling, punctuation and grammar accurately.

Figure 9.4 Student example analysing 'unwanted' existing textile products gathered from home to reuse

Once you have written your specification you will need to 'produce creative and original ideas by generating, developing and communicating designs using appropriate strategies'. This means that you will use your creative skills to design and explore a range of ideas by using the following techniques to help you:

- sketches, drawings, free illustration and diagrams
- annotation and notes which are descriptive and evaluative against your specification criteria
- ICT, CAD and image manipulation software and/or use of digital technologies, photography, etc.
- 2-D and 3-D modelling techniques focusing on specific details and areas of interest. This could involve the making of a toile to test and trial different techniques, skills and processes.

Summarise your thoughts with a **final design idea,** identified and analysed to include reasoned decisions about materials and components.

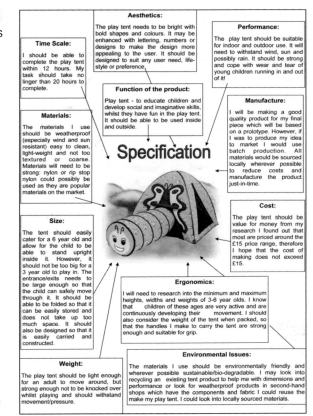

Figure 9.5 Student example of a detailed specification

Figure 9.6 Student examples of design ideas using CAD and free illustration techniques

Figure 9.7 Student example showing exploration into colours, fabrics and components using 2-D modelling techniques

Making (24 marks)

This section involves product planning and realisation, where you will **plan and make** your prototype product. You will be marked according to the way you work and organise yourself. You will need to:

- plan and organise activities:
  - select appropriate materials
  - select hand and machine tools as appropriate to realise the textile product
- work skilfully and safely to assemble, construct and finish materials and components as appropriate

- assess and apply knowledge of the workshop / design studio facilities as appropriate to realise the textile product. **(16 marks)**

This means that you need to select and use tools, equipment and processes safely and effectively. You should list or highlight the tools and equipment you have used within your plan of making. Keep all your pattern pieces to add to this section and make your prototype product as carefully as you can.

Within this section you can also gain marks for 'demonstrating a practical and thorough understanding and ability in solving technical

EXAMINER'S TIPS

- You will need to evidence any changes in use of equipment due to difficulties or problems experienced. This can be done within your plan of making as an action point or written in a separate paragraph/chart.
- Remember, emphasis is placed on the modelling and production of a functional prototype.

problems effectively and efficiently as they arise'. **(4 marks)**

This means you must keep a record of all your problems and explain why these happened and what improvements you made to overcome them.

This section also expects you to 'record key stages involved in the making of the product, provide comprehensive notes and photographic evidence'. **(4 marks)**

This means that you need to produce a detailed real-time record of progress (plan of action) showing the key stages for making your product, using a clear, logical sequence. This could be done as a diary or log or as a video diary.

You must ensure that you use digital photography or video at some point throughout your plan of making to show the key stages of production.

Another important feature to consider in your plan of making is **quality control** to ensure that the product is the best it can be.

Critical evaluation (7 marks)

In this section you are asked to 'critically evaluate the processes involved in making the final prototype product' and 'reflect and suggest modifications to improve the making process'. Chapter 8 in this text book will help you to structure your answer to this section in more detail.

Critical evaluation requires you to be able to:

- **Test your results,** taking into consideration the data collected and the processes you have used. You can do this through the use of a questionnaire targeting your user group. This will help you to suggest modifications to improve the making process. What aspects make the product suitable or unsuitable for the user?

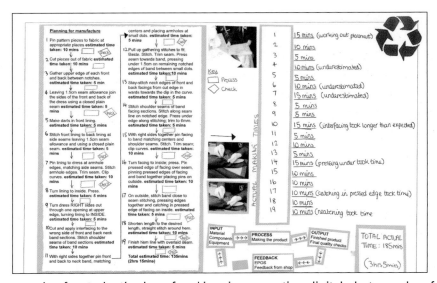

Figure 9.8 An example of a student's plan of making, incorporating digital photographs of key stages of production

- **Predict** – evaluate alternatives to materials, components and processes. This will help you to review how decisions were made (for example, did you use the correct materials?).

- **Evaluate** and monitor your own performance – look at specific setbacks and how you overcame them. Evaluate the importance of each of the key stages of production – research, specification, design, modelling, quality control and making. Comment on the effectiveness of your project management and time planning.

- **Explore** alternative ideas or solutions to problems encountered in the making process (for example, what are the implications of choice of process on moral, ethical and environmental issues?).

- **Modify** – outline realistic and detailed modifications to your work. Use annotated sketches and modelling to support your points. Improvements should be creative and refer to innovation, environmental and sustainability issues.

You need to include a minimum of *two* digital images/photographs of the final product, showing front and back views.

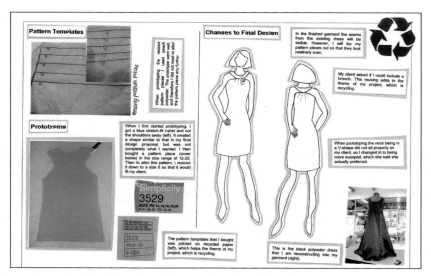

Figure 9.9 Student example showing part of an evaluation of their prototype

CHAPTER 10

UNIT A572: SUSTAINABLE DESIGN

By the end of this section you should have developed a knowledge and understanding of:

- The 6Rs
- Social issues
- Moral issues
- Cultural issues
- Environmental issues
- Design issues

This unit of the GCSE course aims to develop your knowledge and understanding of sustainability, environmental concerns, cultural, moral and social issues.

You will work through this unit in your chosen subject/material area:

You will look at how design and technology have evolved, through analysis of products from the past and the present. You will need to consider how future designs/products will impact on the world in which we live.

By looking at old and new products you will gain awareness and understanding of trends and innovations in design and manufacture, labelling, packaging, and the impact that the design of these products is having on the environment, society and the economy.

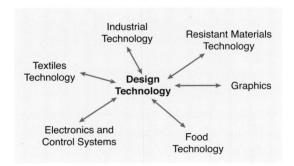

Figure 10.1 Design Technology subject areas

10.1 OVERVIEW OF THE UNIT

Moral, cultural, economic, environmental and sustainability issues are an important part of design and technology. Through this unit you will be able to answer some of the following questions:

- What is meant by a 'product life cycle'?
- Why were certain materials chosen and used?
- What is meant by planned obsolescence?
- What do we mean by the 6Rs?
- What can we do to ensure eventual disposal of products/materials is as eco-friendly as possible?

The assessment of this unit is through an externally set and marked exam.

- You will answer the questions in the exam using your knowledge of your specialist subject/material area.
- This can be taken in either the January or June examination session.
- The unit can be retaken once, with the best result used.
- It represents 20 per cent of a full GCSE qualification or 40 per cent of a short course qualification.
- The maximum mark for the unit is 60.
- The duration of the examination is one hour and the paper is divided into two sections.
- **Section A** will consist of 15 short-answer questions. They will be a mixture of multiple choice, one-word answers and true or false questions. The section will carry 15 marks in total. It is expected that you will spend 15 minutes on this section.

Exemplar question:

Which of the following in **not** a renewable energy resource?

a. water

b. coal

c. wind

d. solar power

- **Section B** consists of three questions which will require you to relate your knowledge and understanding of the '6Rs', materials, processes and the design of products.

Each question will be marked out of 15 marks. The questions may involve sketching, annotation, short sentences or more extended writing. It is expected that you will spend 45 minutes of this section.

Exemplar question:

(a) Identify a product that could be recycled
(b) Explain why you have chosen this product and how it can be recycled   [2]

In section B, your quality of written 'communication' will be assessed as well as your knowledge and understanding. Questions where this is to be done will be marked with an asterisk (*) and will usually ask you to discuss or explain something in detail.

Exemplar question:

* Products become 'obsolete' after a few years. Discuss the difference between fashion and planned obsolescence.   [6]

Make sure you understand the 'command' words that are used in examination papers.

'State ... name ... give'

This requires a specific name of, for example, a piece of equipment, material or process

'Complete'

This requires you to complete, for example, a table, design or drawing.

'Describe'

This requires you to give an idea of, for example, how something works or what is involved in a process.

'Use sketches and notes to ... '

You can use either or both here and it is important that the notes support and expand upon the sketches used.

'Explain'

This requires a detailed account of something including reasons, justifications or comparisons. This type of question will carry two or more marks.

'Discuss'

When you are asked to 'discuss' you must give well-reasoned points and explanations, adding examples to show the examiner what you are thinking. One-word answers, lists or bullet points are not acceptable for this type of question.

EXAMINER'S TIPS

Key skills to achieve high marks in this unit are:

- Think with an open mind about design, be aware of changes that are happening.
- Recall, select, use and communicate knowledge and understanding of concepts, issues and terminology within your material area.
- Record ideas showing design thinking, innovation and flair; this will involve detailed notes and, where appropriate, high-quality sketches and annotated drawings.
- Seek out and use information from existing designers.
- Analyse and evaluate design and production skills and techniques.
- Understand materials, ingredients and components in the context of the chosen product.
- Consider how past and present design technology affects society.
- Demonstrate understanding of the wider effects of design and technology on issues including sustainability, society, the economy and the environment.

10.2 THE 6Rs

RETHINK	How can it do the job better? Is it energy-efficient? Has it been designed for disassembly?
REUSE	Which parts can I use again? Has it another valuable use without processing it?
RECYCLE	How easy is it to take apart? How can the parts be used again? How much energy to reprocess parts?
REPAIR	Which parts can be replaced? Which parts are going to fail? How easy is it to replace parts?
REDUCE	Which parts are not needed? Do we need as much material? Can we simplify the product?
REFUSE	Is it really necessary? Is it going to last? Is it fair trade? Is it unfashionable to be trendy and too costly to be stylish?

Table 10.1 **The 6Rs**

Figure 10.2 Recycling logo

Recycle

Recycling is what we do with the objects we use in our daily lives. Recycling is the conversion of waste products into new materials, to extend the life and usefulness of a product, item or object that seems to have no more purpose or use once it has been finished with or used for its initial purpose. Recycling means reusing a product, but sometimes before a product can be reused it will need to undergo processing or treatment.

There are three main types of recycling:

Primary recycling

The second-hand use of items, whether clothing, electronic or product-based, is a form of primary recycling, as the item is simply being used again. Charity shops stock a large selection of recycled products. Giving items to friends and relatives or selling them on using internet market sites are ways of primary recycling.

Secondary or physical recycling

This is the process in which waste materials are recycled into different types of products.

Figure 10.3 **Second-hand clothing**

The change the product will go through depends on the main fibre or material of the product. Some products can be left to biodegrade before being regenerated into something else. Packaging used for food is often difficult to recycle. However, biodegradable packaging such as 'Potatopak' has been developed.

ACTIVITY

You can research packaging and look at the advantages and disadvantages of this type of packaging. List the materials the packaging is made from and research alternative materials that could be recycled.

Tertiary or chemical recycling

Products are broken down and reformulated. For example, plastic bottles can be recycled into fibres and then respun into polyester to make fleece fabric used for coats and blankets. Car tyres can be reused to make numerous products, for example computer mouse mats.

Recyclable materials include glass, paper,

Figure 10.4 Old tyres

metals, wood, textiles, electronics, tyres, plastics and food wastes. Most, if not all, things can be recycled in some way.

Figure 10.5 Paper being recycled at a waste collection plant

Why recycle?

Everything we dispose of goes somewhere, although once the container or bag of rubbish is out of our hands and out of our house we forget it instantly. Our consumer lifestyle is rapidly filling up rubbish dumps all over the world and, as this happens, concern for the environment grows. When designing and making a new product, designers and manufacturers need to consider how their product can be recycled at the end of its life cycle.

You will need to know the following:

• materials that can be recycled
• products that use recycled materials
• disassembly – reprocessing materials for use in new products.

ACTIVITY

- What does the term recycling mean? Write your own definition.

- List three products that can be recycled.

- Name a material made from recycled products.

Figure 10.6 Recycling plastic, metal, glass and paper

▶ Reuse

Products that can be reused for either the same purpose or a new purpose

Products that are designed to be reused result in less waste; this leads to conservation of materials and resources. Many places around the UK collect unwanted products or repair them for redistribution for the same or a similar end use.

Products that can be adapted to suit an alternative use

Some local areas have set up their own websites and organisations for the reuse of unwanted items. These are run by groups of people who aim to adapt existing products for alternative uses.

▶ Reduce

Life cycle of a product

A new product progresses through a variety of stages from the original idea to its decline, where it might be discontinued or disposed of. You must consider the impact of a product on the environment and its impact on society as a whole. The main stages for consideration are:

- The raw materials – how are they harvested, made?

- The production process – how is the product made?

- Transport and distribution – you need to consider what, how, where and what cost?

- Uses – what are the intended uses of the product? How will it be used by the client or the customer?

- Recycling – how can the product be recycled?

- Care and maintenance – what is needed, how much, and is it environmentally friendly?

- Disposal – the waste from manufacturing or the product itself. Ask yourself the question: is it recyclable or biodegradable?

KEY TERM

LIFE CYCLE means the stages a new product goes through from conception to eventual decomposition.

Eco footprint

This is the term used to refer to the measurement of the impact of our actions on the environment. You as a designer must consider the effect of your product on the environment, from the first stages of your design ideas through to the final making and eventual disposal or recycling of your product. Your footprint involves showing that you have designed the product with the environment in mind and have tried to minimise the damage caused by the various stages throughout your product's life cycle.

Built-in obsolescence

This is where the product has been designed to last a set period of time. The components of the product have been designed by the manufacturer to fail after a certain time limit. The consumer is then under pressure to purchase again. Built-in obsolescence is found in many different products from vehicles to light bulbs, items of clothing to food 'use by' or 'best before' dates. Manufacturers can design the product to become obsolete faster by making the product with cheaper materials which will not last as long as higher-quality components.

Energy and waste of production process

Transportation of products uses large quantities of oil and petrol, which are refined fossil fuels. The consumption of non-renewable energy resources such as coal and oil are causing an energy crisis. These resources will eventually run out. Using non-renewable resources also adds to the world's pollution problem, as products made from oil often take a long time to break down in the environment. 'Green energy' refers to the use of alternative energy sources which are considered environmentally friendly and non-polluting. Renewable and non-polluting sources of energy are:

- wind power

- solar power
- geothermal power
- hydro power
- tidal/wave power.

Figure 10.7 Wind turbines

EXAMINER'S TIPS

The questions shown are open-ended questions and will require your answer to discuss relevant points in the context of your subject/material area.

Use of specialist terms and factual information used appropriately will allow you to score at the higher mark level.

Material waste

We often overlook how much we waste as consumers, whether it be consumable products or power sources such as electricity. Waste management is a growing problem, from chemicals that get into the water system to the materials used in packaging. Switching off our computers or not leaving the television on standby can help us to reduce the energy wasted. Reusing carrier bags or buying locally-made products helps

QUESTIONS

1. Do methods of transportation harm the environment?

2. Are we using too much electricity or too many chemicals, which could harm our environment?

3. What alternative sources of energy are available?

us to reduce material waste and use less resources.

Manufacturers now have to follow guidelines on how to dispose of their waste. Research into effective management of pollution, energy and other material waste is constantly ongoing. You need to be aware of current changes within these areas.

Refuse

Issues relating to sustainable design

Processing, manufacturing, packaging and transporting our products use huge amounts of energy and can create lots of waste. You need to look at the sustainability of a product from an environmental and social viewpoint. How is the product made and can we ensure that no or little harm is done to the environment by this method of manufacture? Sometimes a choice between the performance required of the product and the impact on the environment through its manufacture has to be considered and debated.

Materials we should refuse to use

Why should you refuse to use some products? There are a variety of reasons:

- It may be because the product is made unnecessarily from a manufactured source when a natural source exists.

- It might be because of the toxic chemicals used on the product.

- What about the manufacturing process itself – has it been made in compliance with safety regulations?

- What about the rights of the workers and the conditions they have been working in?

- What about the packaging and transport distances and costs?

You should think about these issues before you accept a product, and above all do not buy it if you do not need it!

Rethink

Within your own lifestyle, and those of others close to you, you need to rethink the way in which you buy products and the energy required to use them. Society is constantly evolving and changing and you can evaluate how you could make a difference.

- How it is possible to approach design problems differently – what ideas can you develop to ensure a difference?

- When reusing an existing product that has become waste is it possible to use the materials or components for another purpose without processing them? What can you design? What could be designed?

Repair

The throwaway society of today means it is quicker, easier and sometimes cheaper to throw something away rather than repair it. You have looked at built-in obsolescence earlier in this chapter, whereby manufacturers encourage consumers to repurchase rather than repair.

- Some products you can repair yourself, some we can take to repair shops.
- Some products are beyond repair or would cost too much to fix.
- Unwanted electronic and electrical

equipment is the fastest growing waste area. Why? It is important for our environment that this attitude is changed. How can you do this?

ACTIVITY

1. In groups discuss what makes you want to buy a product.

2. Discuss and consider what you have bought recently and why. Did you really need it?

10.3 SOCIAL ISSUES

Figure 10.8 Global unity logo

Today we live in a global society. You need to be aware of the ways this can affect the designing of products. Products need to be designed for use by a range of different cultures and nationalities all of whom may have different specific needs. Society today has become multicultural and diverse, so some products may be designed for a specific section of society and others may be universal.

- Consider social development through recognising the need to take into account the views of others when designing and discussing designed products.

- Understand the relationship between man and the general environment.

- Consider the economic development cycle of a range of products and the impact on individuals, societies and countries.

- Keep in mind issues associated with economic development and employment: where a product is made, costs of components, materials, manufacturing process (including labour) and the transportation of the finished product.

- Be aware of the values of society (both your own and others) – why we wear clothes: protection, modesty, adornment. Clothing, for example, has become a way of reflecting our gender, culture and religion. Some items have become unisex and fit across all of society.

Moral issues

Moral issues are concerned with the way in which products are manufactured and the way in which they affect the safety, comfort and wellbeing of people who make them and those who come into contact with the designs/products. Many companies now try and ensure that products are made in the right conditions without exploiting workers, and that they follow a code of practice.

Moral issues that need to be considered are:

- **Moral development**: reflecting on how technology affects the environment and the advantages and disadvantages of new technologies to local and national communities, for example GM foods, production automation, manufacture in developing economies.

- **Conditions of working** within a manufacturing environment: for example, job satisfaction, wages, safety of workplace and workers.

- The **Ethical Trading Initiative** (ETI) which is an alliance of companies, non-governmental organisations (NGOs) and trade union organisations. Their aim is to promote and improve the implementation of regulated codes of practice that set out minimal working requirements. A useful website is: www.ethicaltrade.org. Ethical companies ensure that their employees have basic labour rights and also take care to protect the environment in the production, packaging and distribution of their goods. Ethical companies are often termed 'sweatshop-free'. A sweatshop is a term used to describe a business with poor working conditions.

- **Fairtrade**: the Fairtrade Foundation is the independent non-profit organisation that licenses use of the 'FAIRTRADE Mark' on products in the UK in accordance with internationally agreed Fairtrade standards set by Fairtrade Labelling Organisations

Figure 10.9 FAIRTRADE Mark

International (FLO). The Foundation was established in 1992. The website is www.fairtrade.org.uk

Cultural issues

Culture is about the way that people behave and relate to one another. It is about the way that people live, work and spend their leisure time and it is about their beliefs and aspirations. Many cultures have important traditions that form part of their identity. The use and maintenance of traditional skills and cultural knowledge can have an impact on modern products. Consider how products affect the quality of lives within different cultures. Look at, respond to and value the responses of others to design solutions.

Environmental issues

In a modern, fast-moving society where products are continually being changed, it is important that you keep up to date with various issues. You will need to address the following key areas within your material area:

- Understand and be able to select materials that are both suitable and sustainable.

- Be aware of the disposal and recycling of materials and components and the appropriate methods of manufacture.

- Prepare materials economically, minimising waste and using pre-manufactured standard components.

- Be aware of the need to reduce the common use of environmentally unfriendly chemicals and materials dangerous to the environment, i.e. bleaches, CFCs, toxic materials. The pollution caused by manufacturing can be high, and ways to reduce this are being investigated. It is sometimes necessary to use chemicals and man-made materials which can be harmful to the environment if there are no suitable substitutes.

CFCs

CFCs are one of a group of synthetic substances containing chlorine and bromine, developed in the 1930s. Thought to be safe, non-flammable and non-toxic, they were widely used until the 1980s when it was discovered that they were the main source of harm to the ozone layer.

Carbon footprint

The carbon footprint is a measure of the impact human activities have on the environment in terms of the amount of greenhouse gases produced through the emission of carbon dioxide. The production of carbon dioxide contributes to global warming. The carbon footprint is linked to the ecological (or eco) footprint which can be measured by comparing the impact of energy used through transportation of materials and goods and in manufacture with the effects of using natural resources and renewable resources. The ecological footprint is used to measure our ecological performance; the carbon footprint measures the adverse effect our actions have upon the environment.

Carbon offsetting

This is a method by which people and companies can undertake measures to offset the impact they have on the environment in terms of their carbon footprint. Carbon offsetting involves contributing to the development of more environmentally friendly methods of energy generation, such as the use of renewable sources.

Reforestation

Reforestation is the term used to describe the restocking of existing forests and woodlands. Not only does reforestation replenish a natural resource, but it has other advantages. Trees produce oxygen and also remove carbon dioxide from the atmosphere, offsetting some of the problems caused by

Figure 10.10 Carbon footprint logo

Figure 10.11 Informative labelling is useful to the consumer

the pollution created through manufacturing and transportation.

End-of-life disposal

This issue is linked to the need to dispose of redundant products and their packaging in a safe and environmentally friendly way. The use of recycling labelling for specific packaging is helpful to the consumer when buying products and disposing of them responsibly.

10.4 DESIGN ISSUES

Buying a product can be expensive, so you need to ensure that you have got what you want and that it will benefit you in some way. Research of the product beforehand and analysing the information gathered can help you draw a conclusion, ensuring your choice is successful.

Designers are constantly changing and evolving their work. Sources of inspiration come from all design and technology areas. In all products, new and constantly changing materials are being developed; smart materials in all the subject areas have developed massively over the last few years. To help you keep up to date in these areas it is useful to visit some of the following websites:

- www.voltaicsystems.com: new fabrics made from recycled soda bottles for solar bags
- www.geofabrics.com: fabrics used in landfill, earthworks, railway tracks and drainage

Figure 10.12 European Eco-label

ACTIVITY

1. Identify how good design and product choice improve the quality of life. Create a mind map of your findings.

2. Look at the way companies such as M&S address greener design issues. M&S has an online greener living shop with several fashion collections. Produce a mood board showing one or more of these collections.

3. How do you decide when to update your clothes or other products? Why do you want to buy the latest mobile phone? Do a survey among your friends and make a chart of the reasons given.

- www.seeitsafe.co.uk: hygienic protection against harmful bacteria in a range of products
- http://tinyurl.com/yrmgfs: sustainable fashion – fashion for the future.

Eco–design

Eco-design involves the whole system of looking at an end product from design to finished article in terms of its use of materials and energy. It is the process of designing a product from scratch with the environment in mind and trying to minimise the damage caused to the environment by the product's life cycle.

A designer must think through the following main stages if the product is to be successful and acceptable as eco-designed:

- product planning
- product development
- design process
- functionality
- safety
- ergonomics
- technical issues and requirements
- design aesthetics.

The European Ecolabel is an official label awarded to a product guaranteeing it has fulfilled specific criteria. A product awarded the Ecolabel will have been found to have a smaller environmental impact than other similar products. The Ecolabel is the official sign of environmental quality. It is awarded by independent organisations, and it is valid throughout Europe. The label's criteria aim to limit the environmental impacts of a product over its entire life cycle by looking at such issues as energy and water consumption, waste production and use of renewable resources.

10.5 THE GLOBALISATION OF PRODUCTS

The globalisation of products can mean that products previously only available in a certain country or countries are now available in different countries all around the world, like Levi jeans or Nike trainers. It can also mean that a product of a firm based in one country is manufactured in a different country. This happens because labour can be cheaper in some countries, particularly Third World countries, so the company producing the product can either sell it more cheaply or make a bigger profit. This way of producing goods has advantages and disadvantages for the local population. It can provide jobs and valuable income for people in poor areas, but it can also lead to sweatshops where the workers are exploited and paid very low wages for working long hours. Another moral issue to consider with globalisation is that of harm to the environment and use of fuel resources through transporting goods that could be manufactured locally long distances, particularly by air.

Globalisation of products means that local cultures, ways of doing things and skills can influence the design of the product or lead to new products more suited to local markets. Different cultures have different needs and what is a requirement and need of one culture can be very different from those of another.

Sustainable design is a world issue and a constantly changing one. You should want the world to be a great sustainable place to live in, one that is for you, your friends and relatives and for future generations to come.

A sustainable way of designing can have an impact and positive effect on everyone. As a designer you need to remember and consider the social, economic and environmental implications of your decisions.

ACTIVITY

1. Working in a group; list the advantages and disadvantages you would need to be aware of when having products manufactured abroad. Remember to consider the different materials, the culture and working conditions.

2. Try and list six examples of products that you know have been manufactured abroad.

KEY TERMS

LIFE CYCLE means the stages a new product goes through from conception to eventual decomposition.

REFORESTATION is the term used to describe the restocking of existing forests and woodlands.

SWEATSHOP is a term used to describe a business with poor working conditions.

CULTURE is the way that people behave and relate to one another; it is the way that people live, work and spend their leisure time. It also reflects people's beliefs and aspirations.

CFCs are one of a group of synthetic substances containing chlorine and bromine, developed in the 1930s. Thought to be safe, non-flammable and non-toxic, they were widely used until the 1980s when it was discovered that they were the main source of harm to the ozone layer.

CHAPTER 11

UNIT A573: MAKING QUALITY PRODUCTS

In this section you will learn about:

- The assessment requirements of this unit
- The evidence to be presented within the portfolio
- Options for presenting the portfolio

In this controlled assessment unit you are expected to further develop the skills and abilities gained from completing Unit A571 in order to design and make a fully functioning quality product. You will be able to select one of the themes published by OCR to use as a starting point for this unit. It is important to remember that the themes will be reviewed every two years and that you need to select a different theme from the one followed in Unit A571. Your teacher will be able to provide you with a list of the current themes.

11.1 STRUCTURE

Unit A573 makes up 30 per cent of the total GCSE marks for the full course. It is a 20-hour controlled assessment task which has 60 marks in total. The work associated with this unit must be completed under your teacher's supervision. However, some of your research work and testing of the product, by their very nature, may take place outside school under limited supervision. You must reference the source of any information you use within your portfolio.

Submission of controlled assessment work

Your work can be submitted on paper or in electronic format but not a mixture of both.

Work submitted on paper

A contents page with a numbering system should be included to aid organisation. All your work should be on the same size paper but it does not matter what size paper you use. You can produce your work by hand or by using ICT.

Work submitted electronically

Your work will need to be organised in folders so that the evidence can be accessed easily by a teacher or moderator. This structure is commonly known as a folder tree. There should be a top-level folder detailing your centre number, candidate number, surname and forename, together with the unit code A573. The next folder down should be called 'Home Page', which will be an index of all your work. The evidence for each section of the work should then be contained in a separate folder, containing appropriately named files. These files can be from the Microsoft® Office suite, movie, audio, graphics, animation, structured mark-up or text formats. PowerPoint® is ideal for producing electronic portfolios.

11.2 UNIT REQUIREMENTS

This unit will require you to produce the following:

- A production plan.
- A number of concise worksheets showing design development and modelling, which may include ICT used to support the design process.
- A product capable of evaluation.
- A minimum of *two* digital images/photographs of the final product, showing front and back views of the product in use.
- Digital images/photographs of any models or mock-ups used when designing, modelling or testing.
- A completed OCR cover sheet.

You will be expected to include *only* the following areas within your controlled assessment portfolio:

Figure 11.1

- designing for a need
- working with tools and equipment
- making a textile product
- an evaluation of the product.

11.3 GUIDANCE POINTS

In order to skilfully design, make and evaluate your textile product and complete your supporting portfolio, you need to read the following guidance points based on an OCR published theme.

Theme: Contemporary design

Starting point / outline: 'Create a textile bag in the style of a favourite fibre artist or contemporary designer for an identified user group.'

▶ Designing (16 marks)

This section is divided into two parts; the first will expect you to:

'Demonstrate an appropriate and considered response to a brief and produce a detailed specification for a textile product as a result of analysis.' **(4 marks)**

This means that you will need to write down the theme and your starting point clearly and use these to identify 'areas of action' through a **mind map** or **thought diagram**. This will help you to think about the key stages needed to complete your portfolio successfully.

Research

Research the theme 'contemporary design' – consider the meaning of contemporary design. Produce information sheets showing the ways contemporary design has been influenced in the home, in fashion and accessories. For example, the influence of Indian textiles on contemporary design; high street retail outlets such as Avant Clothing, who specialise in modern contemporary work, and designers like Divya Thakur, who is well known for her Indian-influenced contemporary design shunya bags. Explore

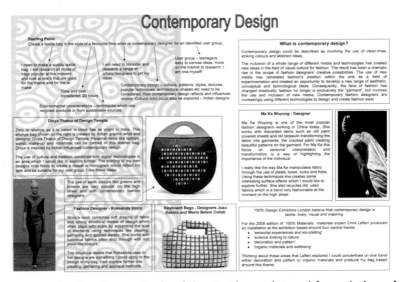

Figure 11.2 Student example showing analysis of the starting point and formulation of a brief

the links between bags and design in detail and analyse your findings linked to specific user groups.

To find out the needs of different user groups, you could complete a questionnaire designed to analyse the choice factor, for example cost benefits, range of sizes, colours, styles, preferred designers, fabrics and production techniques.

Design brief

At the end of this section you should be able to show that you have considered and researched the starting point in order to write a design brief. An example of a brief written in response to the theme 'Contemporary design' could be:

'I shall design and create a textile bag suitable for a teenager, and influenced by the style and works of Divya Thakur.'

EXAMINER'S TIPS

- Remember to use appropriate textile examples. Keep a note of websites and resources you have used (research and quotes should be 'acknowledged for source' next to each piece when written, *not* presented as a bibliography at the end of the portfolio).
- Present your information with flair and creativity.
- Remember, a questionnaire can incorporate visual stimuli as well as written questions to analyse consumer choice.

Specification

You will need to collate and analyse your findings based around the principles of good design. Conclude your results from your questionnaire and research, discussing the advantages and disadvantages of your findings. You then need to write a detailed **specification.** Chapter 1 will give you more guidance with how to write your specification.

EXAMINER'S TIP

- Make sure that you justify each specification point and consider environmental and sustainability issues.

The second part of this section asks you to:

'Produce a wide range of creative and original ideas by generating, developing and communicating designs using appropriate strategies.' **(12 marks)**

This means that you will use your creative skills to design and explore a wide range of design ideas by using the following techniques to help you:

- sketches, drawings, free illustration and diagrams
- annotation and notes which are descriptive and evaluative against your specification criteria
- ICT, CAD and image manipulation software and/or use of digital technologies, photography, etc.
- 2-D and 3-D modelling techniques and the use of CAM, focusing on specific details and areas of interest. This could involve

the making of a toile to test and trial different techniques, skills and processes to aid product development.

EXAMINER'S TIP

- A range of creative ideas (6-8 marks) will cover 3-4 design ideas, whereas a wide range (9-12 marks) will consider 5-6 original and different design ideas.

Summarise your thoughts with a **final design idea** identified and analysed to include reasoned decisions about materials and components.

▶ Making (36 marks)

This section involves product planning and realisation, where you will **plan and make** your quality textile product. You will be marked according to the way you work and organise yourself. This means that:

- you need to plan and organise activities:
- select appropriate materials
- select hand and machine tools as appropriate to realise the textile product
- you need to work skilfully and safely to assemble, construct and finish materials and components as appropriate when making quality textile products.
- you need to assess and apply knowledge of the workshop/design studio facilities as appropriate to realise the textile product
- the product will be completed to a high standard and will fully meet the requirements of the product specification. **(24 marks)**

This means that you need to select and use tools, equipment and processes safely and effectively. You will also need to show that you can work skilfully and independently to shape, form and finish materials and assemble components.

It is enough to list or highlight the tools and equipment you have used within your plan of making. Keep all your pattern pieces to add to this section and make your product as carefully as you can. It is important that this is a finished quality textile product that can be evaluated.

Within this section you can also gain marks for 'demonstrating a practical and thorough understanding and ability in solving technical problems effectively and efficiently as they arise'. **(6 marks)**.

This means you must keep a record of all your problems and explain why these

Figure 11.3 Student example showing exploration into colours and fabrics using 2-D modelling techniques

EXAMINER'S TIPS

- You will need to evidence any changes in use of equipment due to difficulties or problems experienced. This can be done within your plan of making as an action point or written in a separate paragraph/chart.
- Remember, it is better to make one well-made quality textile product which has used materials and components appropriately than to make a complicated product which looks rushed in appearance and finish.

happened and what improvements you made to overcome them.

You are also expected to 'record key stages involved in the making of the product, provide comprehensive notes and photographic evidence'. **(6 marks)**.

This means that you need to produce a detailed real-time record of progress (plan of making)

showing the key stages in making your product, using a clear, logical sequence. This could be done as a diary or log or as a video diary.

You must ensure that you use digital photography or video at some point throughout your plan of making to show the key stages of production.

Another important feature to consider in your plan of making is **quality control** to ensure that the product is the best it can be.

Critical evaluation (8 marks)

In this section you are asked to 'critically evaluate the finished product against the specification'.

This means that you need to evaluate each of your specification points in detail – use examples from your portfolio to help justify your reasoning and consider factors like cost, theme, user need, purpose of the product, environmental, moral and cultural issues, suitability of techniques and processes used, industrial production methods, ease of

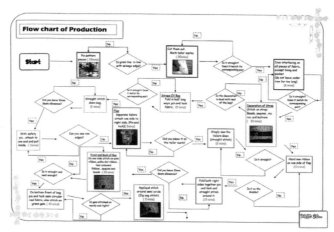

Figure 11.4 An example of a student's plan of making, incorporating digital photographs of key stages of production

making, suitability of materials and components used, etc.

You will also need to 'undertake detailed testing; present meaningful conclusions leading to proposals for modifications to improve the product'.

This means you will need to:

- Take into consideration the data collected and the processes you have used. You can do this through the use of a questionnaire targeting your user group. This will help you to suggest modifications to improve the making process. What aspects make the product suitable or unsuitable for the user?
- Include a minimum of *two* digital images/photographs of the final product in use.
- Explore alternative ideas or solutions to problems encountered in the making process – what are the implications of choice of process for moral, ethical and environmental issues?
- Outline realistic and detailed modifications to your work. Use annotated sketches and modelling to support your points. Improvements should be creative and refer to innovation, environmental and sustainability issues.

Chapter 8 will help you to structure your answer to this section in more detail.

You will be able to gain extra marks for this section by showing that you have considered how to present the information gathered in a structured way and your spelling, punctuation and grammar are accurate.

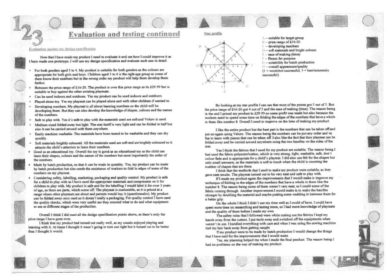

Figure 11.5 Student's evaluation piece for a textile product showing critical evaluation against the specification

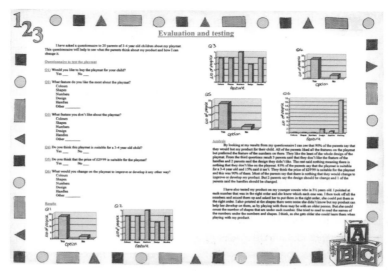

Figure 11.6 Example of student's work showing testing of the final textile product

CHAPTER 12

UNIT A574: TECHNICAL ASPECTS OF DESIGNING AND MAKING

In this section you will learn about:

- The format of the examination for unit A574
- The types of questions to expect
- How to prepare for the examination
- Examination technique

Only students following the GCSE full course take this examination. The unit A574 examination is 1 hour 15 minutes in length and represents 20 per cent of your final GCSE marks. It is externally marked.

12.1 THE FORMAT OF THE EXAMINATION

The examination has five questions. This means you have fifteen minutes for each question, so there is no need to rush. You need to answer all of the questions; there is no choice of question. The maximum mark available is 60 and each question is marked out of 12. The marks you get from this paper will be added to the marks you achieve in the A572 examination and the marks for your two controlled assessment pieces to give you your final GCSE grade.

Your teacher will decide when to enter you for the examination. It can be taken in January or June. The examination is divided into two sections:

- **Section A** has three questions based on the technical aspects of designing and making. Each question will be subdivided into shorter questions.

- **Section B** has two questions. One of these questions will ask you to design something. The second question in Section B will be about sustainability and human use. You will be asked to think about how textile products affect the world around us, the benefits of them and the problems they can cause.

12.2 TYPES OF QUESTIONS

The specification for this examination states that you will be asked to:

- recall, select and communicate your knowledge and understanding in Design Technology including its wider effects
- analyse and evaluate products including their design and production.

Recall means remember, so you will need to remember things you have been taught during the course. You may have done some written work, or some practical work that will help you in the examination. Don't forget, when working on the controlled assessment pieces you will have learnt about designing, planning, how to do practical tasks and much more. You should remember and use this information when answering questions in the examination. You will need to 'select' or choose the information needed to answer the question.

The questions in the examination are about the 'technical aspects of designing and making'. This can cover a wide range of topics, and some, but not all, are listed below.

- Where fibres come from and how they are made into yarns and fabrics. This includes 'smart and modern' materials.
- The performance characteristics of fibres, yarns, and fabrics and why they are used for particular products.
- The finishes that can be applied to fabrics to improve their appearance and performance characteristics.
- Explaining why particular fabrics are used for specific products.

- Explaining how the production, use and disposal of textile items affect the environment.
- Knowing about components needed to make products, how to choose them and how to attach them.
- The tools and equipment used to make products – how to choose them and use them safely.
- The processes used to make products – using paper patterns, marking fabric, disposal of fullness, joining fabrics, neatening edges, pressing and finishing methods, quality control in these processes.
- How to decorate fabrics – with dyes and with fabric and thread.
- Industrial techniques used to mass-produce items as well as those used for 'one-off' production.
- How computers are used when designing and making.
- Health and safety.
- Quality control.
- The 6Rs.
- Product analysis.

One of the best ways to find out about the format of the questions on the examination paper is to look at past papers. Your teacher will be able to provide these for you to look at and answer. You will need to be able to write down your answers and use diagrams and sketches to explain techniques used when making textile products.

In Section A each question is divided into sections, and each will contain some harder parts as well as some easier ones. You will need to write down your answers, either in sentences or phrases, possibly even a list. Some questions will ask you to draw diagrams and sketches to explain things as well as using words. Some questions may show you a product and then ask you questions about it. This is known as product analysis.

> # Example question

Figure 12.1 shows a sports bag.

Figure 12.1

(a) Figure 12.2 shows one of the pattern pieces used to make the bag.

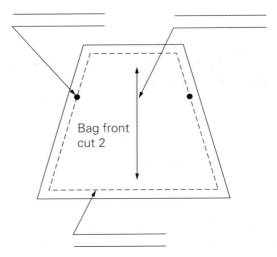

Figure 12.2

Copy out the diagram and complete the labels to show the meaning of the pattern symbols.   [3]

(b) The bags are to be manufactured using the **'batch' production system**.

Explain **two** advantages of using the **'batch' production system**.

(c) Evaluate the effectiveness of using ICT to develop the pattern pieces for the bag.   [5]

[Total: 12]

Mark scheme

Question Number	Section A	
	Answer	Max Mark
(a)	Figure 12.1 shows a sports bag. The diagram below shows one of the pattern pieces used to make the bag. Complete the labels on the diagram to show the meaning of the pattern symbols. One mark for each correct answer: • Straight grain arrow/grain line • Dot/tailor tack/balance mark • Stitching line/seam line/fitting line	[3]
(b)	The bags are to be manufactured using the 'batch' production system. Explain two advantages of using the 'batch' production system. Any two points explained, one mark for a shallow explanation, two if detailed: • Cheap – fabric and components can be bought in bulk, saving money • Quality product made – workers repeat tasks, so become skilled and faster • Colour changes are easy to effect – little tooling/machinery to change, only need to change colour of thread • Flexible to deal with orders for different colours/quantities/demand • Quick/efficient – team workers, large number of people working together, repetition of task increases speed • More items made at the same time – increases profits • All products the same (size) – improves quality/consistency, so customer benefits	[2+2]
(c)	Evaluate the effectiveness of using ICT to develop the pattern pieces for the bag. Shows limited understanding of the uses of ICT and how effective ICT could be to develop the pattern pieces. [0-2 marks] Shows some understanding of how effective ICT could be to develop the pattern pieces with some analysis of the issues involved. Basic conclusion may be drawn. [3-4 marks] Shows detailed understanding of how effective ICT could be to develop the pattern pieces and analyses most of the issues involved. Appropriate conclusions are drawn. [5 marks]	
	Evaluation may include reference to: • Quicker/saves time • More accurate/less human error • Can be used to generate a lay plan • Can be stored on disk, saving place • Can be easily adapted/changed/graded/modified • Can be emailed to clients/other manufacturers • Can be downloaded directly to cutter • Can reduce costs by reducing the workforce	[5]

Table 12.1

Remember, one of the questions in Section B will ask you to design something. You will need to come up with a couple of ideas and then develop one of them into a final design. You will need to annotate your ideas; in other words, label them. The labels will need to clarify what the design will look like and give information such as colours, fabrics, components and processes to be used to make the product. You will need to be able to explain how and why your design answers the question.

▌ Example question

1. A company wishes to include a wall hanging for a child's room in its product range.

The specification for the product is to:

• hold a range of small toys;

• have educational value;

• be environmentally friendly to produce.

(a) Use sketches and notes to show your initial ideas.   [4]

(b) Show your final design idea.

Annotate your sketch to show all important design and construction details.   [8]

The second question in Section B will be about sustainability and human use. You will be asked to think about how textile products affect the world around us, the benefits of them and the problems they can cause. Some of the work done for the unit A572 examination will help here.

▌ Example Question

Consumers are increasingly aware of the need to protect and preserve the environment.

2 Consumers often tire of textile products before they reach the end of their useful life.

Discuss how such textile products can be given a new lease of life.   [6]

▌ Mark scheme

Section B		
Question Number	Answer	Max Mark
1(a)	A company wishes to include a wall hanging for a child's room in its product range. The specification for the product is to: • hold a range of small toys; • have educational value; • be environmentally friendly to produce. Use sketches and notes to show your initial ideas. Marks allocated as follows: • 1 mark if only one sketch with no accompanying notes • 2 marks for a sketched solution with notes • 3 marks if more than one sketch with notes • 4 marks for a range of solutions with notes relating back to the specification	[4]

Table 12.2

Section B		
Question Number	Answer	Max Mark
1(b)	In the space below, show your final design idea. Annotate your sketch to show all important design and construction details. A maximum of 8 marks to be allocated as detailed below: • Colour indicated [1] • Measurements given [1] • Fastenings show [up to 2 marks] • Suitable decoration/motif/logo [1] • Pockets in a range of sizes [1] • Fabrics suggested (not fibres) [1] • Construction details given, seams, hems, finishing methods [up to 2 marks] • Decorative techniques given, appliqué, screen printing, machine stitching etc. [up to 2 marks] • Educational value explained [up to 2 marks] • Environmental issues explained [up to 2 marks] • More than one sketch included – detail of a specific part	[8]
2	Consumers often tire of textile products before they reach the end of their useful life. Discuss how such textile products can be given a new lease of life. Level 1 (0-2 marks) Basic discussion, showing limited understanding of how textile products can be given a new lease of life. There will be little or no use of specialist terms. Answers may be ambiguous or disorganised. Errors of grammar, punctuation and spelling may be intrusive. Level 2 (3-4 marks) Adequate discussion, showing some understanding of how textile products can be given a new lease of life. There will be some use of specialist terms, although these may not always be used appropriately. The information will be presented for the most part in a structured format. There may be occasional errors in spelling, grammar and punctuation. Level 3 (5-6 marks) Thorough discussion, showing detailed understanding of how textile products can be given a new lease of life. Specialist terms will be used appropriately and correctly. The information will be presented in a structured format. The candidate can demonstrate the accurate use of spelling, punctuation and grammar. Discussion may include: • Potential for giving products to charity shop/organisation – for clothing, household goods and fabric toys. • Pass clothing to younger children/other families if outgrown. • Dye it to make it more appealing.	

	Section B	
Question Number	Answer	Max Mark
	• Add decoration to it, e.g. appliqué, hand stitching, beading, lace, ribbon. • Cut it up and make it into something new – e.g. bedding. • Use the fabric for patchwork or appliqué. • Re-fashion it – e.g. make a bag from a pair of jeans. • Pass it on to a manufacturer who can reclaim the fibres, e.g. wool. • Take off pre-manufactured components such as buttons and zips which can be re-used for other items. • Use for cleaning cloths.	[6]

12.3 PREPARING FOR THE EXAMINATION

Throughout this book, at the end of a section of information, there is a series of activities. These are designed to help you gather and learn the information you need to answer the questions in this examination. Keep all of the completed tasks in a Research Folder and use the folder to revise from before taking the examination. The better job you make of completing the tasks, the better the information you will have to revise from.

Your teacher may also set 'focused practical tasks'. These are fairly short, practical activities designed to help you develop a range of technical skills and gain knowledge of materials and processes. These should be kept to revise from.

You should also complete 'design and make' assignments that include activities relating to sustainability of products and resources as well as industrial practices.

This will also link in with Unit A572, which focuses on sustainable design. These will involve practical activities as well as design work and research.

Product analysis is a good way of learning about products, what they are made from and how they are made. It is good to carry out product analysis on a range of textile items and keep the information in your research folder.

Your teacher may take you to particular technology innovation centres, museums and industries. This is a good way of carrying out research and it is important to gather as much information as possible on these visits. The internet is obviously useful for carrying out research, but it is important to be selective about the information you gather and include in your research folder.

12.4 TAKING THE EXAMINATION

It is important to revise thoroughly before the examination. Use your Research Folder, but don't forget about the work you have done for your controlled assessment pieces. If you are thoroughly prepared, you will find the examination less stressful.

Take at least two pens into the examination, black or blue. You will also find drawing and colouring equipment useful for the design question. Remember, felt-tip pens can go through the paper to the other side, so coloured pencils are best. If you don't have colouring equipment, you can label the design to show the colours.

The examination will be marked 'online', which means the paper will be scanned. It is very important to write or draw your answers in the places indicated so that the examiner can find them. Your writing and drawings need to be dark enough to show up when the paper has been scanned.

Remember, there is no need to rush, you have fifteen minutes for each question. Read each question through before you start. This will help you to understand what the question is about, and may give you some clues.

Look at the number of marks available for each section of the question, and the amount of space given for the answer. This will give you an idea of how much detail to include, and how many points to include in your answers. Words such as 'give', 'state' or 'name' usually mean one-word answers are enough, whereas words such as 'explain', 'analyse' or 'justify' indicate a more structured, detailed answer is needed. If the question asks for a diagram, or for an explanation using notes and diagrams, make sure you include them.

There is a mix of easier and harder questions throughout the paper. If you find a question difficult, try your best to answer it. Marks are not taken off for wrong answers and it is better to attempt the question rather than leave it blank. Carry on working through the paper, you may find some easier questions later on.

Remember, the examiner needs to be able to read your answers, so write clearly. Examiners are quite good at working out what you mean, even if the spelling is not quite right.

Try to use the correct technical terms and vocabulary, particularly when writing longer answers.

If you have time left at the end of the examination, check through what you have done. Fill in any blank spaces with educated guesses based on common sense. Remember, the examination is your chance to show what you know, so do your best!

INDEX

TEXTILES
13